Achieving a Sustainable Global Energy System

ESRI STUDIES SERIES ON THE ENVIRONMENT

In April 2000 the Japanese government launched a series of comprehensive, interdisciplinary and international research projects called 'the Millennium Projects' and as part of this initiative the Economic and Social Research Institute (ESRI) of the Cabinet Office of Japan initiated a two year project entitled 'A Study on Sustainable Economic and Social Structures in the 21st Century', which focuses on ageing and environmental problems in the Japanese and international context.

The *ESRI Studies Series on the Environment* provides a forum for the publication of a limited number of books, which are the result of this international collaboration, on four main issues: research on solid waste management; the analysis of waste recycling and the conservation of resources and energy; research on the compatibility of environmental protection and macroeconomic policy and the analysis of problems related to climate change. The series is invaluable to students and scholars of environment and ecology as well as consultants and practitioners involved in environmental policy making.

Titles in the series include:

Firms, Governments and Climate Policy
Incentive-based Policies for Long-term Climate Change
Edited by Carlo Carraro and Christian Egenhofer

Eco-Efficiency, Regulation and Sustainable Business
Towards a Governance Structure for Sustainable Development
Edited by Raimund Bleischwitz and Peter Hennicke

Achieving a Sustainable Global Energy System
Identifying Possibilities Using Long-Term Energy Scenarios
Leo Schrattenholzer, Asami Miketa, Keywan Riahi and Richard Alexander Roehrl

Achieving a Sustainable Global Energy System

Identifying Possibilities Using Long-Term Energy Scenarios

Leo Schrattenholzer

International Institute for Applied Systems Analysis (IIASA), Austria

Asami Miketa

International Institute for Applied Systems Analysis (IIASA), Austria

Keywan Riahi

International Institute for Applied Systems Analysis (IIASA), Austria

Richard Alexander Roehrl

United Nations (UN/ESA/ESCAP), Thailand

With a contribution from Manfred Strubegger, Gerhard Totschnig and Bing Zhu

ESRI STUDIES SERIES ON THE ENVIRONMENT

IN ASSOCIATION WITH THE INTERNATIONAL INSTITUTE FOR APPLIED SYSTEMS ANALYSIS (IIASA)

Edward Elgar

Cheltenham, UK • Northampton, MA, USA

Published by
Edward Elgar Publishing Limited
Glensanda House
Montpellier Parade
Cheltenham
Glos GL50 1UA
UK

Edward Elgar Publishing, Inc.
136 West Street
Suite 202
Northampton
Massachusetts 01060
USA

A catalogue record for this book
is available from the British Library

Library of Congress Cataloguing in Publication Data
Achieving a sustainable global energy system : identifying possibilities using long-term energy scenarios / Leo Schrattenholzer . . . [et al.] ; with a contribution from Manfred Strubegger, Gerhard Totschnig and Bing Zhu.
p. cm. – (ERSI studies series on the environment)
1. Energy policy. 2. Energy development. 3. Energy conservation. 4. Sustainable development. I. Schrattenholzer, Leo. II. Series.

HD 9502.A2A325 2005
333.79—dc22

2004050641

ISBN 1 84376 923 9

Printed and bound in Great Britain by MPG Books Ltd, Bodmin, Cornwall

The International Institute for Applied Systems Analysis
is an interdisciplinary, nongovernmental research institution founded in 1972 by leading scientific organizations in 12 countries. Situated near Vienna, in the center of Europe, IIASA has been producing valuable scientific research on economic, technological, and environmental issues for over three decades.

IIASA was one of the first international institutes to systematically study global issues of environment, technology, and development. IIASA's Governing Council states that the Institute's goal is: *to conduct international and interdisciplinary scientific studies to provide timely and relevant information and options, addressing critical issues of global environmental, economic, and social change, for the benefit of the public, the scientific community, and national and international institutions.* Research is organized around three central themes:

- Energy and Technology;
- Environment and Natural Resources;
- Population and Society.

The Institute now has National Member Organizations in the following countries:

Austria
The Austrian Academy of Sciences

China
National Natural Science Foundation of China

Czech Republic
The Academy of Sciences of the Czech Republic

Egypt
Academy of Scientific Research and Technology (ASRT)

Estonia
Estonian Association for Systems Analysis

Finland
The Finnish Committee for IIASA

Germany
The Association for the Advancement of IIASA

Hungary
The Hungarian Committee for Applied Systems Analysis

Japan
The Japan Committee for IIASA

Netherlands
The Netherlands Organization for Scientific Research (NWO)

Norway
The Research Council of Norway

Poland
The Polish Academy of Sciences

Russian Federation
The Russian Academy of Sciences

Sweden
The Swedish Research Council for Environment, Agricultural Sciences and Spatial Planning (FORMAS)

Ukraine
The Ukrainian Academy of Sciences

United States of America
The American Academy of Arts and Sciences

Contents

Foreword

This book reports on work done by IIASA-ECS in recent years, in particular the group's contribution to the 'Collaboration Project for the Economic Planning Agency Japan (EPA)'. During this project, also known as the 'Millennium Project', an international group of researchers worked in two areas, aging and environmental issues. IIASA-ECS contributed to the environment area by presenting and analysing long-term energy–economy–environment scenarios in general and sustainable development scenarios in particular. We are very grateful to the Japanese government for the sponsorship of this work. In particular, we thank Dr Hiromi Kato of the Economic and Social Research Institute, Cabinet Office, Government of Japan. We also thank our colleagues from the other collaborating groups, in particular Professor Carlo Carraro of the University of Venice, Professor Frank Convery of University College, Dublin, and Christian Egenhofer from CEPS, Brussels for helpful comments and suggestions.

The basis of the analyses done in this study has been derived from the experience and earlier work done by IIASA-ECS scientists for the IIASA-WEC studies as well as the IPCC Special Report on Emissions Scenarios. We would like to acknowledge Nebojša Nakićenović, who led ECS until January 2000, as well as Arnulf Grübler, Sabine Messner, Manfred Strubegger, Hans Holger Rogner, Alan McDonald and Andrei Gritsevskyi, as well as the contributions of all the authors who have been actively involved in these studies.

Preface

The world at the beginning of the 21st century must place the highest priority on constructing a sustainable socioeconomic system that can cope with the rapid ageing of populations in developed countries and with the limited environmental resources available in both developed and developing countries. At first glance, the problems of ageing and the environment may seem to be quite separate issues. However, they have a common feature: they both deal with intergenerational problems. The essence of the ageing problem is how to find effective ways for a smaller working generation to support a larger, ageing generation. The crux of the environmental problem is to find a feasible way to leave environmental resources to future generations. Moreover, in terms of consumption, slower population growth may slow consumption and help environmental problems. On the other hand, a rapidly ageing society may use more energy-intensive technology to compensate for the inevitable labour shortage, and deteriorate the natural environment by doing so.

Today, these concerns are highly applicable in Japan. The pressure created by the rapid ageing of the Japanese population is becoming acute; Japan must construct a sustainable society that does not create intergenerational inequity or deteriorate the public welfare. At the same time, Japan cannot deplete its environmental resources and energy, which would leave future generations with an unbearably heavy burden.

The government of Japan has recognized the vital importance of both problems. To explore and implement solutions for this difficult task, in April 2000 former Prime Minister Keizo Obuchi launched several comprehensive and interdisciplinary research projects that he called the 'Millennium Project'. As a part of these projects, the Economic and Social Research Institute (ESRI), Cabinet Office, Government of Japan, initiated a two-year project entitled 'A Study on Sustainable Economic and Social Structure in the 21st Century' in April 2000. While the Millennium Project covers a wide range of topics and a wide range of disciplines such as natural science and technological innovation, the project conducted by ESRI places major emphasis on social science. While taking into account technological innovation and feasibility, it focuses on ageing and environmental problems. It aims to design a desirable socioeconomic structure under the pressure of an ageing population and environmental constraints by identifying the necessary policy tools to attain stable and sustainable growth.

This project is being implemented with close collaboration among Japanese as well as foreign scholars and research institutes. Besides Japanese scholars and institutes, foreign participants have been involved from, among other countries, the USA, the UK, Norway, Austria, Italy, Australia, Korea and Thailand. In all, there are ten countries and 30 working groups.

In this project, ESRI explores optimal solutions to problems in social science terms. After taking into account the political and social constraints we face, and after alignment and coordination with the results of the studies, it sketches an ideal design and examines the possible direction of future research. This project came to an end in March 2001. It resolved many theoretical and empirical issues, but has created new debates. Twice a year, all the participants in the project, along with invited others, meet to discuss the results of the research. Regrettably, it has not been possible to reproduce the fruitful discussion in the present volume.

Overall, the papers presented in the project were extremely challenging, and covered a wide range of topics. In the near future we strongly hope we will have a chance to discuss the research once more from a common standpoint. The result of this research is published by Edward Elgar Publishing Ltd as part of an ESRI study series, available to policy makers, academics and business people with a keen interest in these subjects. The series on environmental problems covers climate change, sources of energy and technology, and environmental and employment policy. Unfortunately, because of space limitations, we regret that we are able to publish only selected papers from the total research effort. The research papers to be published were selected by the Editorial Board members. We would like to acknowledge the ceaseless efforts of the members of ESRI throughout the project period, especially those of the Department of Administration Affairs. Last but not least, we would like to thank Dymphna Evans from Edward Elgar Publishing.

Yutaka Kosai, President, ESRI

1. Sustainable development and climate change

Since the early 1990s, the concept of sustainable development has been receiving considerable and increasing attention by scientists and policy makers alike. It has become common to look separately at three parts of the general concept. These are social, economic and environmental sustainability. As to the global environment, climate change is the issue that dominates policy making and analysis alike, and many groups analysing climate change embed this issue in the overall framework of sustainable development.

Energy use is central to climate change, but also to sustainability in general. Addressing both goals at the same time leads to the formulation and analysis of strategies that lead to environmentally compatible and sustainable energy systems, which is the main theme of this book. Sustainability is rarely studied by one discipline alone; nonetheless, interdisciplinary studies of sustainable development have a focus. The study presented in this book was conducted by the Environmentally Compatible Energy Strategies (ECS) Program at the International Institute for Applied Systems Analysis (IIASA). It focuses on environmentally compatible long-term developments of the global energy–economy system, but also includes aspects of economic and social sustainability. It is a matter of course, but it shall be explicitly emphasized that the sustainability of the global energy system as presented here can only be partial. The global system as a whole could still be unsustainable in aspects that are beyond the system boundaries considered here.

Even with this focus on the energy system, the field of sustainable development is vast and characterized by high uncertainty. In fact, the intrinsic uncertainty surrounding the long-term development of the global energy–economy–environment system[1] is so high that the conclusiveness of studies such as ours is the recurring subject of questions asked during its frequent presentations by the authors. One of the most common methods of addressing such uncertainty is the use of scenarios, and this is also the approach used during this study. We have analysed a large number of scenarios to illustrate the difference between those that meet a set of criteria for sustainable development and those that do not.

We have extracted common characteristic features of the sustainable-development scenarios, but we have identified quite a variety of socio-economic and environmental developments that are consistent with sustainable development. In this book we aim at portraying this variety, but also at presenting one sustainable-development scenario in more detail from a policy-making perspective.

As to the conclusiveness of our results, we think that it is more indicative than cogent. A possible conclusion that we think would be adequate to the subject matter could be: 'I like the sustainable-development scenario described here, and I will therefore contribute to achieving it.' Since it is primarily policy makers who are in a position to contribute, we describe our study in policy-relevant terms.

1.1 SUSTAINABLE DEVELOPMENT

Many definitions of 'sustainable development' have been proposed. One of the least controversial definitions has been formulated by the Brundtland Commission as 'development that meets the needs of the present without compromising the ability of future generations to meet their own needs' (WCED, 1987). This definition gives a generally accepted basic characterization of the concept, and broadly defines a policy direction. However, for this concept to become operational in policy making, we need to have a more concrete concept of 'sustainable development'.

Many authors have undertaken initial steps towards concretization of the general concept by defining measurable indicators of sustainable development (compare, for example, Pearce *et al.*, 1996; Klaassen and Opschoor, 1991; Tietenberg, 2000), but this is not a straightforward task. Three issues are important sources of conceptual difficulties in defining sustainable development. In our view, they are, first, completeness of the set of indicators; second, the measurability of indicators; and, third, their commensurability. The last point is another formulation of the well-known 'apples and oranges' metaphor. Here it says that pairs of indicators might be measured in different units, which in many cases means that possible trade-offs between improving either one or the other indicator can be assessed differently by different proponents of sustainable development.

Addressing the completeness of indicators, usually three major components are distinguished: economic, environmental and social sustainability. In our opinion, these three are ranked from most straightforward (economic) to most difficult to quantify (social). Beginning with the simpler task, we discuss economic sustainability first. The modern concept of *economic sustainability* underscores the sustainability of the economic benefit from natural assets. The rationale behind this idea contends that the flow of economic benefit of natural assets should be preserved because it should be shared between the current and future generations. A typical argument along this line can be found in El Serafy (1989), who states that not all revenues from selling natural resources should be treated as current income that is available for consumption. This idea upholds the possibility of a substitution between man-made assets and natural assets.

This possibility of substitution leads to the distinction between weak economic sustainability and strong economic sustainability. The concept of weak economic sustainability permits reinvesting the revenues from selling natural resources in man-made capital as long as doing so yields at least as much output as the forgone natural resources. In contrast to this, the concept of strong economic sustainability is based on the premise that natural and man-made capital offer only limited substitution possibilities and thus require the separate preservation of natural resources and other capital. According to this notion, sustainability is defined as non-declining value of the remaining stock of natural capital (Tietenberg, 2000). Strong sustainability thus permits the use of fossil resources only if the value of the remaining resources does not decrease. Moreover, the strong version of economic sustainability requires the separate preservation of each category of critical assets (for example, manufactured, natural, socio-cultural and human capital), assuming they are complements rather than substitutes.

The notion of *environmental sustainability* goes even further and requires the maintenance of the 'physical property of the environment'. This view requires the preservation of the ecological function of the environment, which is defined in terms of scientific knowledge on ecological property of natural assets. This requirement still permits human consumption of natural assets (such as clean air), provided that the ecological function of the environment recovers in the near term (Munasinghe, 2000).

In addition to these – at least in principle – measurable indicators of sustainability, other, non-quantifiable indicators must be considered. These include cultural assets (such as historical buildings), nuisance (such as noise), traffic flow, and others. Some of these can be understood as part of *social sustainability*, although the most prominent indicator of social sustainability is social equity (ibid.). These criteria are evaluated differently by different people, which often makes the definition of appropriate indicators of social sustainability a contentious issue.

To our knowledge, and so far, none of the attempts to define sustainable development in quantitative terms has received a broad acceptance. The major reason for this lack of agreement comes from the fact that such attempts usually focus on establishing sustainable development as a one-dimensional objective. Still, without quantification, there is no hope of making the concept of sustainability operational. We therefore ventured one step towards the practical applicability of the sustainability concept by conceiving quantitative criteria that permit a classification of *existing* energy–economy–environment (E3) scenarios as 'sustainable' or 'non-sustainable'. We chose a multiple-criteria approach, which is likely to be less controversial than a one-dimensional criterion. The advantage of having a multi-criteria approach is its flexibility with respect to emphasizing one view or the other. We shall summarize our concept in section 1.3, but before, we want to define what we understand under scenarios.

1.2 SCENARIOS

In this book we address sustainable development by analysing long-term global E3 scenarios. We begin with the definitions of the term 'scenario' and of the sustainable-development (SD) scenario. For the purpose of this book, we define a scenario as *a consistent and complete description of (the development of) a system*. In our case, completeness is defined by the formulation of the E3 model MESSAGE, which has been used to formulate the central SD scenario analysed in this book. MESSAGE is described in non-technical terms in Chapter 2, and in more complete and technical terms in the appendix.

It is important to distinguish a scenario from forecasts and, even more clearly, from predictions, the most important difference being that forecasts and, to an even higher degree, predictions are meant to portray particularly likely future developments. In contrast, scenarios often include elements that may not be considered the most likely development. This is particularly true for sustainable-development scenarios, which have important normative (prescriptive) elements. Rather, sustainable-development scenarios are meant to enrich the reader's imagination by portraying the possible and, in some instances, by exploring the limits of the plausible. This endeavour involves a considerable amount of subjective judgment, but a major motivation for writing this book was to describe the path from past developments over assumptions to sustainable-development scenarios in a transparent way. We do this in particular by illustrating this process by describing one particular sustainable-development scenario in some detail. Our intention is to provide readers with a basis for assessing that scenario using their own judgment, helping them to form their own opinion on the plausibility of our assumptions in particular and of sustainable-development E3 scenarios in general.

The second definition required for the understanding of this book is that of sustainable-development scenarios. We provide this definition in the following section.

1.3 SUSTAINABLE-DEVELOPMENT SCENARIOS

For practical purposes, we have adopted a working definition of sustainable development that has been inspired by the 'Brundtland spirit' referred to above, but that is also sufficiently quantitative to serve as a tool for classifying long-term energy-economic scenarios. This means that the quantities used for the working definition are either parameters or outputs of the models applied in the scenario development.

For a working definition we refer to Klaassen *et al.* (2002). All scenarios that satisfy the following four criteria will be referred to as sustainable-development scenarios.

(1) Economic growth (GDP/capita) is sustained throughout the whole time horizon.
(2) Socioeconomic inequity among regions (that is, intragenerational equity), expressed as the world-regional differences of GDP per capita, is reduced significantly over the 21st century, in the sense that, by 2100, the per capita income ratios between all world regions are reduced to ratios close to those prevailing between OECD countries today.
(3) Long-term environmental stress is mitigated significantly. In particular, carbon dioxide emissions at the end of the 21st century are approximately at or below today's emissions. Other greenhouse gas (GHG) emissions may increase, but total radiative forcing, which determines global warming, is on a path to long-term stabilization. Other long-term environmental stress to be mitigated includes impacts on land use, e.g., desertification. Short- to medium-term environmental stress (e.g., acidification) may not exceed critical loads that threaten long-term habitat well-being.
(4) The reserves-to-production (R/P) ratios of exhaustible primary energy carriers do not decrease substantially from today's levels. This criterion reflects the principle of intergenerational equity.

The authors believe that this working definition is close to that of weak economic sustainability, complemented by environmental and social constraints. Furthermore, this study is taking a global and long-term perspective. This means, for instance, that a scenario that may appear unsustainable in the near term may still follow a sustainable path in the longer-term future.

1.4 AUDIENCE AND OBJECTIVES

The main aim of this book is to specify and analyse a set of possible circumstances that is consistent with a sustainable path of future developments of the global energy–economy–environment system. These circumstances primarily describe technological progress, and our analysis aims at identifying the evolution in technologies that will be needed if sustainable development of the global energy sector is to be achieved. The idea is to outline a picture of a future world with sustained global economic growth, a movement towards a fairer distribution of products and services and an energy production that becomes increasingly environmentally compatible.

The analysis addresses itself to the interested public as well as to policy makers. The policy relevance of a global long-term study inevitably is more general than decision-aiding analyses of the near-term future. In our opinion, this study primarily suggests targets for technology developments. These targets may appear ambitious, but the possible reward for success is a sustainable development of the global energy sector. Although we do not provide detailed recipes – one reason being that there are several strategic options for doing so – we discuss and categorize policy options in general terms. The available policy options include market-based instruments (taxes and subsidies), financial instruments, public procurement, environmental treaties and the support of energy-related research and development (R&D). The actual choices will largely depend on the political environment in any given country, but the authors are convinced that energy-related R&D is the most important. For the sustainable-development (SD) scenario described in detail in Chapter 5, we will therefore attempt an approximate estimate of total R&D support required to lead to the technological progress as specified in this scenario.

For those who may wish to replicate our scenarios, we also include some methodological insights obtained during our work. However, to increase reading efficiency, we made an attempt to separate clearly the methodological parts from the rest of the subject matter. Most importantly, we have included a detailed technical description of the MESSAGE model in a separate appendix.

1.5 STRUCTURE OF THIS BOOK

The material in this book is structured as follows. Chapter 2 introduces the main methods used in the study. It provides a non-technical overview of the MESSAGE model and explains the process leading to the formulation of long-term E3 scenarios. An overview of the development of the global E3 system during the 20th century provides a frame of reference for the more detailed descriptions of future scenarios throughout this book.

Chapter 3 reviews a large number of E3 scenarios and categorizes them into three groups: high-impact scenarios, greenhouse gas (GHG) mitigation scenarios, and sustainable-development (SD) scenarios. The purpose is first, to quantify the ranges of important variables determining a scenario and, second, to characterize better SD scenarios as opposed to other long-term E3 scenarios.

Chapter 4 introduces the concept of technology clusters. We have extended the common concept of technology clusters to clusters that are defined, not only with reference to technical criteria, but also in policy-relevant terms such as public acceptance and market success. This extended concept is useful for simplifying the complex structure of the global energy system dynamics, in order not only to 'see the trees, but also the forest'.

Chapter 5 describes one particular sustainable-development (SD) scenario in detail. This is directly compared to a non-sustainable scenario that is similar to the SD scenario in many respects, in particular in terms of the assumed global economic growth. The scenario descriptions will focus on the aspect of energy technology change since, at least in our scenario world, technology change is one of the key drivers of sustained economic growth.

Chapter 6 summarizes the main policy messages of this book, closes with an overview of related work of the IIASA-ECS Program, and offers some thoughts on potentially useful interaction between research and policy making.

NOTE

1. Throughout this book, we shall use the abbreviation E3 for energy–economy–environment. In addition, tons always means metric tons.

REFERENCES

El Serafy, E. (1989), 'The proper calculation of income from depletable natural resources', in Y. Ahmad, S. El Serafy and E. Lutz (eds), *Environmental Accounting for Sustainable Development*, A UNEP–World Bank symposium, Washington, DC: World Bank, pp. 10–18.

Klaassen, G. and J.B. Opschoor (1991), 'Economics of sustainability or the sustainability of economics: different paradigms', *Ecological Economics*, **4**, 93–115.

Klaassen, G., A. Miketa, K. Riahi and L. Schrattenholzer (2002), 'Targeting technological progress towards sustainable development', *Energy and Environment*, **13** (4/5), 553–78.

Munasinghe, M. (2000), 'The nexus of climate change and sustainable development: Applying the sustainomics transdisciplinary meta-framework', in F. Lo, H. Tokuda and N.S. Cooray (eds), *The Sustainable Future of the Global System III*, Proceedings of the International Conference on Sustainable Future of Global System, 24–5 May 2000, Tokyo, Japan: UNU/IAS and OECD.

Pearce, D., K. Hamilton and G. Atkinson (1996), 'Measuring sustainable development: progress on indicators', *Environment and Development Economics*, **1**(1), 85–101.

Tietenberg, T. (2000), *Environmental and Natural Resource Economics*, Reading, MA: Addison-Wesley.

WCED (World Commission on Environment and Development) (1987), *Our Common Future*, Oxford: Oxford University Press.

2. Methodology

This chapter presents the motivation for choosing the particular methods used in our study. After this, we describe, in aggregate terms, how the IIASA-ECS scenarios were built. The description of the scenario building includes an overview of the development of the global E3 (energy–economy–environment) system in the course of the 20th century. We then proceed to give, in non-technical language, a macroscopic description of models and concepts used for the scenario formulation. Together with the models we also describe, in general terms, how the so-called 'driving forces' define scenarios. This way, we give readers an idea of the respective importance of the variables that shape scenarios and, at the same time, an approximate idea about the sensitivity of the results.

2.1 WHY SCENARIO ANALYSIS?

Why do we use scenarios to address the uncertainty surrounding the future development of the global E3 system? And why do we not use stochastic optimization, for instance? Before attempting to answer these questions, we want to define the term 'scenario'. For the purposes of this book, we want a scenario to be understood as an internally consistent and reproducible image of the future. Scenarios are therefore neither predictions nor forecasts. The most important difference between forecasts and scenarios is that scenarios do not necessarily aspire to maximize the likelihood of their occurrence. One prominent kind of scenarios that many would argue are not the most likely to materialize is the class of sustainable-development scenarios. Their main purpose is to specify a set of *possible* circumstances that is consistent with a sustainable path of future developments of the global E3 system. Such scenarios (and others that serve a similar purpose) belong to the class of so-called 'prescriptive' or 'normative' scenarios. In contrast, scenarios that describe the consequence of assuming alternative future states of the world, which usually are meant to be particularly likely, are called 'descriptive'. The distinction between normative and descriptive is conceptual, and borderline cases exist that could be classified either way.

We use scenarios and not stochastic optimization to analyse the uncertainties surrounding the future development of the global E3 system because we think that this is the most suggestive way to present a range of possibilities. Moreover, the distributions of the uncertainties involved – let alone their parameters – are not known. An adequate stochastic treatment of the uncertain variables would therefore have to include a sensitivity analysis with respect to all plausible probability distributions. In our opinion, this would be a hopeless task, not only owing to the large number of such distributions, but also because of the level of abstraction at which this task would have to be performed.

Apart from these methodological considerations, we think that the scenario approach to public policy making is as adequate as the approach used by individuals for their day-to-day private decision making. Take as an example the question whether to take an umbrella with you for the day. One plausible way to find an answer to this question would be to imagine one scenario with and one without rain. And, while imagining these two scenarios alone does not solve the decision problem, it is obvious that they are a useful basis for decision making even without knowing their probabilities of occurrence.

In principle, scenarios are reproducible by anyone who can use the models that were employed in generating them. The consistency of scenarios not only makes sure that there are no hidden contradictions, it also guarantees that, within the chosen boundaries, each scenario is a complete description of the system studied.

2.2 BUILDING LONG-TERM E3 SCENARIOS

2.2.1 The Global Economy–Energy–Environment System in the 20th Century

The basis for future scenarios is laid by past developments. As we are going to present scenarios of the 21st century, we shall first present an overview of the development of the global energy system in the 20th century. For a systematic view of the energy system and its environmental impact in the past and in the future it is useful to disaggregate total emissions into components. Following Kaya (1990), total carbon emissions can be represented by the following identity:

$$CO_2 = POP * \frac{GDP}{POP} * \frac{TPE}{GDP} * \frac{CO_2}{TPE} \tag{2.1}$$

where

CO_2 = carbon emissions,
POP = population,
GDP = gross domestic product,
TPE = total primary energy demand.

In this equation, total primary-energy demand (the result of multiplying the first three terms on the right-hand side of the equation) is conceptualized as the product of population, a measure of welfare (GDP per capita) and the primary-energy intensity of the economy. Emissions of carbon dioxide, the main energy-related greenhouse gas, can thus be thought of as the product of primary-energy demand and its carbon intensity. Using this concept lays the ground for using results of three scientific fields (demography, economics and engineering) to explain and to project energy demand and its aggregate environmental impact. We shall now look at the past trends of each of the variables in turn.

Primary-energy consumption
Total global primary-energy consumption, including all sources of commercial energy and fuel wood, has grown at an average annual rate of approximately 2 per cent per year for more than one century (Watson *et al.*, 1996). This growth corresponds to a doubling of consumption every 35 years. Including the non-commercial use of fuel wood when measuring total primary energy is important for the assessment of the development of the energy intensity of GDP. If non-commercial fuels are omitted, the average energy intensity can show potentially misleading rises over time.

Population
During the 20th century, global population increased from 1.6 to some 6 billion, corresponding to an average annual growth of 1.3 per cent. (See, for example, Grübler and Nakićenović, 1994.) Currently, the world's population is increasing at about 2 per cent per year. While population growth is slowing down in developed countries with comparatively high per capita income, the population in most of the developing countries is still growing at a relatively high pace.

Economic growth
For reasons of data availability, long-term economic growth is best observed in industrialized countries. According to Maddison (1989), the average per capita GDP in 32 industrialized countries increased from 841 US$ in 1900 to 3678 US$ (in constant 1980 prices) in 1987; that is, at an

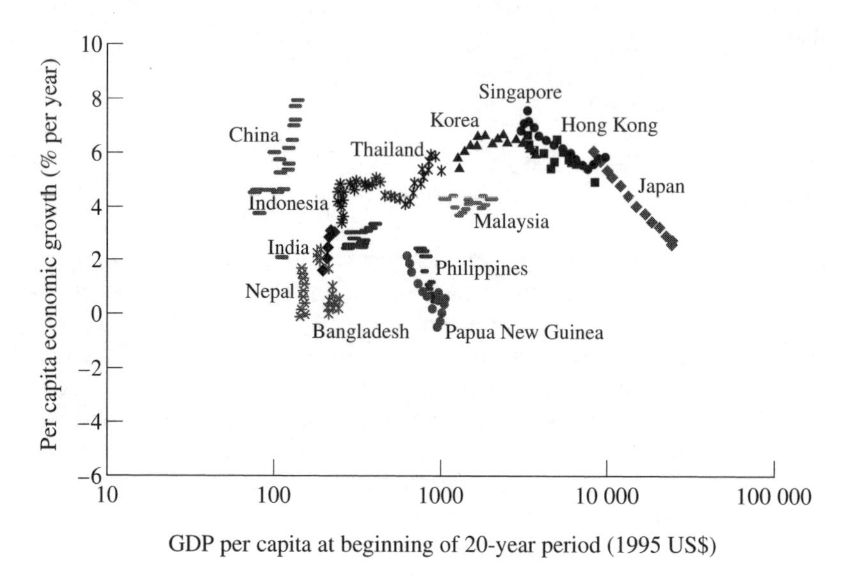

Source: World Bank (2000).

Figure 2.1 Long-term (20-year) average annual per capita economic growth rates in Asian economies, 1960–97, and GDP per capita at the beginning of any of these 20-year periods

average annual rate of 1.7 per cent. Populations in these 32 countries grew at the same annual 1.3 per cent as global population, so that total GDP in these 32 countries grew at an average of 3 per cent per year.

A summary of economic growth rates in Asian economies between 1960 and 1997 is shown in Figure 2.1. The figure shows annual rates of per capita economic growth in Asian economies as a function of GDP per capita. We can see that, with a few exceptions (China on the high side; Bangladesh, the Philippines and Papua New Guinea on the low side), per capita GDP growth rates in Asia have followed a distinct 'inverse U'-shaped pattern. Relations like this, which extend over a range of two orders of magnitude of the independent variable (GDP per capita), in our opinion make for a particularly plausible argument to use it as a reference for long-term projections. We shall therefore take up this point again further (see pages 61–4).

Energy intensity of GDP

From the numbers presented so far we can derive a long-term trend of primary energy intensity reduction of GDP of 1 per cent per year. This is the result of 3 per cent economic growth and 2 per cent growth of total

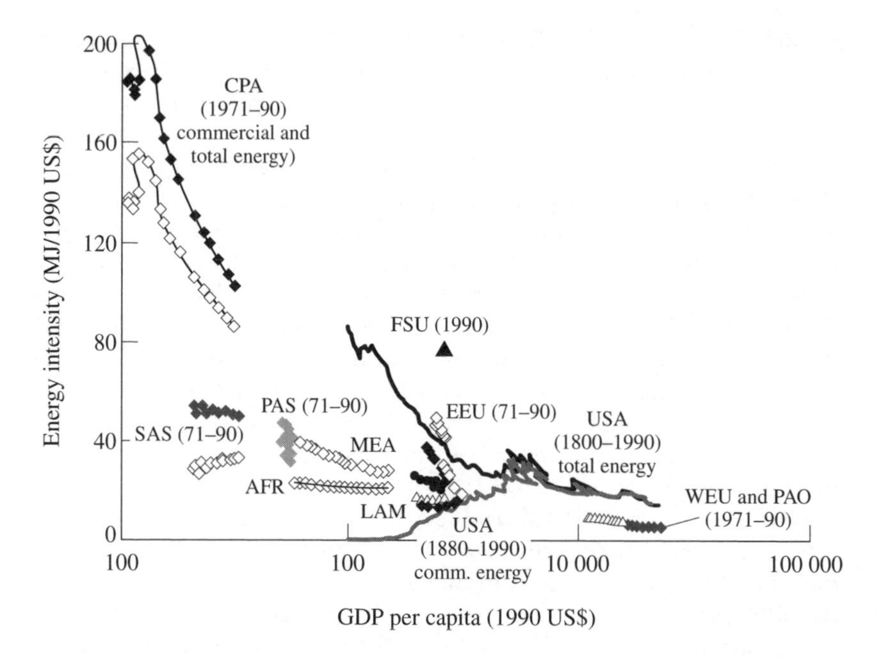

Note: The developments for the USA and CPA are shown for total and commercial energy, respectively. For an explanation of regional abbreviations, see the appendix. MJ=megajoule. For acronyms, see page 174.

Source: Nakićenović *et al.*, 1998 (adapted).

Figure 2.2 Primary energy per GDP (energy intensity) as a function of GDP per capita for 11 world regions

primary-energy demand. A more detailed picture of the development of energy intensities is given in Figure 2.2.

The general picture emerging from Figure 2.2 is one of ever-decreasing energy intensity with significant differences between countries. Here it is important to note that we include non-commercial energy in our analysis. Other studies, for example, Goldemberg *et al.* (1988), do not include non-commercial energy in their calculations of energy intensity and therefore find peaks such as those shown for the USA region in Figure 2.2.

Decarbonization

A decline of the carbon intensity of primary energy has been termed 'decarbonization' (Kanoh, 1992). Since 1860, the carbon intensity of primary energy supply has decreased at an average annual rate of 0.3 per cent

(Nakićenović *et al.*, 1993). Taking this rate together with long-term growth rates of primary energy consumption and GDP, we find that the carbon intensity of GDP has decreased at an average annual rate of 1.3 per cent.

We have just described the past 100 years of the development of the global energy system in terms of what is often referred to as 'slow variables'. We therefore argue that these variables are also the most important descriptors of the future development over a time period of equal length, and we thus consider the projection of these slow variables the key to generating plausible long-term scenarios. Still, even slow variables are not predictable, but for us they serve as baselines relative to which future developments are quantified. Before quantification, however, we describe basic scenario characteristics in qualitative terms. This description is also known as the 'storyline' of a scenario, and we shall describe the path from past development to quantitative scenario specifications via storylines in the following subsection.

What about surprises? This is a very plausible question and one that is often asked, but as intuitively clear as the question appears at first sight, the difficulty is to give it a precise meaning. Is it the low-probability range of distributions of future events? Or is it a factor not considered in the analysis that will play a decisive role in a way that renders the scenarios useless? As to low-probability events, we think that, the more detailed the description of the global E3 system is, the more it will be likely to encounter one surprise of this kind. In an analysis as aggregated as ours, we would argue that the most important low-probability events would be large deviations of the development of slow variables from past trends. But this is exactly the point addressed by scenario analysis, which aims at covering wide ranges around past trends by a number of scenarios. An important illustration of this strategy is the example of the special report on emission scenarios of the intergovernmental panel on climate change (IPCC-SRES) process. One of the first steps in generating IPCC emission scenarios was to collect scenarios from the published literature in the field, thus mapping the range of possibilities considered plausible by the authors.

As to surprises in the form of new and decisive factors, we think that the logic of our scenarios would make them most sensitive to such surprises in the technological area. They would be of the kind that significantly reduces demand or increases supply of cheap energy. Addressing such possible surprises in the technology supply field, we have included all kinds of technologies that have at least demonstrated their feasibility at the laboratory or demonstration stage (for example, hydrogen production with solar energy via the thermal splitting of water, or carbon capture and sequestration in underground reservoirs). We did not consider, however, radically new technological options such as nuclear fusion or so-called 'geo-engineering options' (such as huge solar power satellites systems), since with today's limited knowledge it is inherently unpredictable whether these options can be turned into vital technologies even in the long term. In addition, the development of these technologies from the first steps of the invention process, to demonstration projects, to significant market shares requires too long on average, compared to the time horizon of the scenarios analysed in this book (O'Neill *et al.*, 2003).

2.2.2 Storylines

In the field of long-term scenarios, a storyline is a qualitative description aiming at directing the development path of the slow variables of a scenario, usually relative to their past trends. An example of an element of a storyline of an E3 scenario is 'high economic growth', which means that economic growth to be assumed in a scenario should be on the high side of past trends. Such qualitative characterizations included in storylines are later quantified and transformed into model input numbers. The most important of these 'driving variables' are GDP and energy intensity of GDP, which are combined to determine energy demands.

The projection of aggregate energy intensity in a scenario depends on general technology development and on economic structural change. It is commonly assumed that higher economic growth means faster technological progress and more rapid structural change towards less energy-intensive economic sectors. This is the reason why we think that higher economic growth favours sustainable development.

2.2.3 Preparing Inputs for the MESSAGE Model

At IIASA-ECS, the main tool for the consistent description of long-term E3 scenarios is the energy supply model MESSAGE (Model of Energy Supply Strategy Alternatives and their General Environmental impact). For translating the qualitative assumptions described in the storylines into quantitative model inputs, we use the so-called 'Scenario Generator' (SG). The SG is a conceptually simple spreadsheet model including a database with world–regional information on the slow variables described in the previous subsection. The user of the SG can use built-in econometric tools to calculate past trends from these data. On the basis of these trends, the user can generate time series with additional semi-quantitative commands such as 'higher', 'lower', 'asymptotically approach a value of x' and others. Described in functional terms, the SG converts storylines into MESSAGE inputs from past data, while applying users' judgment. A more detailed description of the SG is provided in the appendix.

But the SG does not produce *all* MESSAGE inputs. On technology development, for instance, MESSAGE requires specific information on the availability (in time) and performance of energy conversion technologies, which cannot be extracted from the SG. The most important of these, specific cost and environmental impact of a technology, must be provided by the MESSAGE user, of course in consistency with the spirit of the specific storyline.

An important concept used in specifying technology development in our E3 scenarios is that of technological learning. According to that concept, technological progress proceeds in a regular fashion as a function of cumulative experience with that technology. For example, the specific investment costs of many energy conversion technologies have been shown to follow nicely what is called 'learning curves'. According to such learning curves, technology cost decreases by a constant (learning) rate each time the cumulative production of that technology doubles.[1] A storyline that specifies fast technological progress can therefore be translated into technological progress that includes learning rates at the higher end of those observed in the past.

Although we think that the success of technology development, measured as specific cost and environmental impact of a technology, depends decisively on energy policy, in particular the support of energy technology research and development, our models do not include formal equations describing such a relation in quantitative terms. The main reason for this omission is that relations describing the impact of R&D on technological performance in quantitative terms do not appear to have a sound empirical basis. Moreover, experimental formulations of this dependence are highly non-linear and therefore computationally difficult to handle. Nonetheless, first experiments with models that optimize R&D expenditures on energy technologies (Miketa and Schrattenholzer, 2004) have yielded indicative results, which we have used in side calculations outside the MESSAGE optimization to estimate the order of magnitude of R&D expenditures that might lead to the technological progress assumed for the sustainable-development scenario described in section 5.3 of this book.

2.2.4 The MESSAGE Model

At IIASA-ECS, the main tool for the consistent description of long-term E3 scenarios is the energy supply model MESSAGE (Model of Energy Supply Strategy Alternatives and their General Environmental impact). The present version of MESSAGE is the result of continuous model development and refinement at IIASA since the 1970s. Recent widely visible applications of the model include the formulation of emission scenarios contributing to IPCC's Special Report on Emission Scenarios (Nakićenović and Swart, 2000) and of stabilization scenarios (Riahi and Roehrl, 2000) for IPCC's Third Assessment Report (Metz *et al.*, 2001).

MESSAGE is a dynamic optimization (cost minimization) model for describing the long-term evolvement of the global energy supply system and its environmental impact. The constraints of the model concern primary-energy resource availabilities, the evolvement of energy conversion technologies and a set of useful-energy demands in seven categories.[2] A detailed description of the MESSAGE model and its most important input parameters is given in the appendix.

For the understanding of the main part of this book it appears sufficient to think of MESSAGE representing a 'Reference Energy System'; that is, all flows from primary-energy extraction to end use via one or more stages of conversion by energy technologies. The model is solved by using commercial optimization software. Finding a model solution, the first task of the software is to determine all flows from primary-energy supply to useful-energy demand that are possible (feasible) within the constraints. Among all feasible flows, the one that incurs minimum discounted costs is identified as the MESSAGE result (which constitutes a scenario).

We have now described the Reference Energy System. Let us now briefly describe the primary-energy side of MESSAGE. We begin with geological resources. The most common concept used in the analysis of geological resources and reserves is the so-called 'McKelvey diagram' (see Figure 2.3), which classifies occurrences of geological resources according to two criteria, economic feasibility and geological assurance. The words used to describe different attributes of these two criteria are not always the same. In Figure 2.3 we follow Rogner (1997).

Note that both criteria of Figure 2.3 depend on technological progress. In the case of drilling for crude oil, the example of North Sea oil shows that the economic feasibility of production can increase through progress in drilling techniques, and geological assurance of any occurrence of minerals can be increased by progress in exploration techniques.

Higher category indices in Figure 2.3 mean higher specific resource costs, and the availability of hydrocarbon and nuclear resources in IIASA scenarios is defined by including all categories up to a given index, which depends on the storyline of that scenario. For the numerical values used to quantify these assumptions, see pages 117–21.

Total amounts of primary energy from renewable sources have, for practical purposes, no a priori cumulative constraints. Owing to their intermittent and diffuse occurrence, MESSAGE includes constraints on their

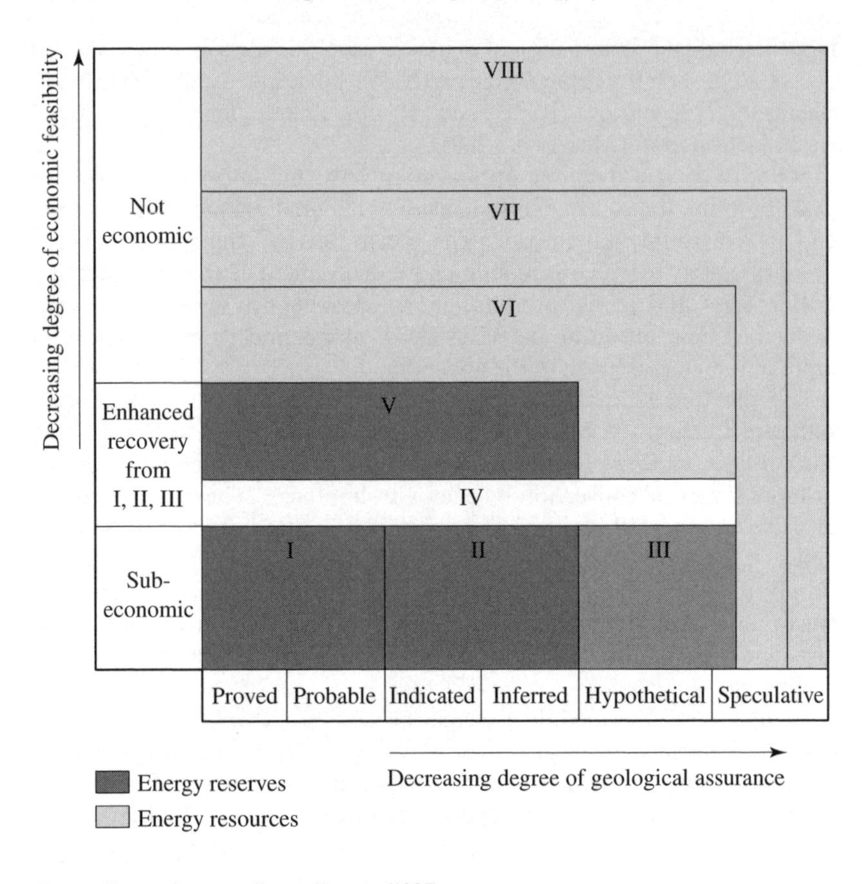

Note: Categories according to Rogner (1997).

Figure 2.3 The classification of energy reserves and resources, McKelvey diagram

annual availability, however. In addition, the model includes constraints on the speed of build-up of capacities that harvest renewable energy.

NOTES

1. For a survey including 42 learning rates of energy technologies, see McDonald and Schrattenholzer (2001).
2. The demand categories are industry thermal, industry non-thermal, residential/commercial thermal, residential/commercial non-thermal, feedstocks, non-commercial and transport.

REFERENCES

Goldemberg, J., T.B. Johansson, A.K. Reddy and R.H. Williams (1988), *Energy for a Sustainable World*, New Delhi: Wiley Eastern.

Grübler, A. and N. Nakićenović (1994), *International Burden Sharing in Greenhouse Gas Reduction*, RR-94-9, Laxenburg, Austria: International Institute for Applied Systems Analysis.

Kanoh, T. (1992), 'Toward dematerialization and decarbonization', *Science and Sustainability, Selected Papers on IIASA's 20th anniversary*, International Institute for Applied Systems Analysis, Laxenburg, Austria.

Kaya, Y. (1990), 'Impact of carbon dioxide emission control on GNP growth: Interpretation of proposed scenarios', paper presented to the IPCC Energy and Industry Subgroup, Response Strategies Working Group, Paris (mimeo).

Maddison, A. (1989), *The World Economy in the 20th Century*, Development Centre Studies, Paris: OECD.

McDonald, A. and L. Schrattenholzer (2001), 'Learning rates for energy technologies', *Energy Policy*, **29**(4), 255–61.

Metz, B., O. Davidson, R. Swart and J. Pan (eds) (2001), *Climate Change 2001: Mitigation*, contribution of Working Group III to the Third Assessment Report of the Intergovernmental Panel on Climate Change, Cambridge, UK: Cambridge University Press.

Miketa, A. and L. Schrattenholzer (2004), 'Experiments with a methodology to model the role of R&D expenditures in energy technology learning processes; first results', *Energy Policy*, **32**(15), 1679–92.

Nakićenović, N. and R. Swart (eds) (2000), *Emissions Scenarios*, Special Report of the Intergovernmental Panel on Climate Change, Cambridge, UK: Cambridge University Press.

Nakićenović, N., A. Grübler and A. McDonald (eds) (1998), *Global Energy Perspectives*, Cambridge: Cambridge University Press.

Nakićenović, N., A. Grübler, A. Inaba, S. Messner, S. Nilsson *et al.* (1993), 'Long-term strategies for mitigating global warming', *Energy*, **18**(5), 401–609.

O'Neill, B., A. Grübler, N. Nakićenović, M. Obersteiner, K. Riahi, L. Schrattenholzer and F. Toth (2003), 'Planning for future energy resources', letter to *Science,* **300**(5619), 581–2, 25 April 2003.

Riahi, K. and A.R. Roehrl (2000), 'Energy technology strategies for carbon dioxide mitigation and sustainable development', *Environmental Economics and Policy Studies*, **3**(2), 89–123.

Rogner, H.H. (1997), 'An assessment of world hydrocarbon resources', *Annual Review of Energy Environment*, **22**, 217–62.

Watson, R., M.C. Zinyowera and R. Moss (eds) (1996), *Climate Change 1995: Impacts, Adaptations and Mitigation of Climate Change: Scientific Analyses*, Contribution of Working Group II to the Second Assessment Report of the Intergovernmental Panel on Climate Change, Cambridge, UK: Cambridge University Press.

World Bank (2000), *World Development Indicators 2000*, Washington, DC: World Bank.

3. Energy–economy–environment scenarios at IIASA-ECS

In this chapter we describe general characteristics of energy–economy–environment scenarios. We characterize three groups of scenarios (high-impact, mitigation and sustainable-development). We then characterize SD scenarios in more detail by comparing ranges of key variables (driving forces and results) of SD scenarios with ranges of the same variables chosen from the IPCC-SRES database of scenarios.

3.1 A COMPREHENSIVE COLLECTION OF ENERGY–ECONOMY–ENVIRONMENT SCENARIOS

Soon after the emergence of the first global long-term energy scenarios, efforts were initiated to compare the results of such scenarios and to learn from their differences. Examples of these efforts include the Energy Modelling Forum,[1] founded in 1976, and the International Energy Workshop, founded in 1981 (Schrattenholzer, 1999). In the course of time, energy scenarios more and more gave way to E3 scenarios, and the efforts to compare their results and to compile them in one place were ever increasing.

One of the latest results in this respect is the database established during the work on IPCC's Special Report on Emission Scenarios (SRES) (Nakićenović and Swart, 2000). This database is therefore also known as the SRES database (Morita and Lee, 1998). It includes the results of some 400 E3 scenarios, which are described in terms of the most important variables characterizing the long-term development of the E3 system either globally or for major world regions. These variables include population, economic growth, energy demand, carbon emissions and others. Although not all scenarios in the database report on all variables, the scenarios included can be regarded as representative of the range of possibilities regarded as plausible by the global modelling community. We will therefore use this database as a frame of reference for the presentation of IIASA scenarios.

The range of opinions held on future values of decisive variables of the E3 system reflects and quantifies the uncertainty surrounding the evolvement of these variables over the course of the 21st century. To what extent the ranges and frequency distributions quantify the uncertainty in any reliable way (that is, to what extent they can be interpreted as probabilities) is open to speculation and individual judgment. In this book we are not going to interpret frequencies as probabilities, and the use of statistical indicators (medians, variances, correlation coefficients and others) are meant in a purely descriptive sense.

3.2 EXPLORING THE RANGES

A common way of graphically presenting values and ranges of a number of variables is a regular polygon and (zero-based) axes between the centre and each vertex. For summarizing the global scenarios of the SRES database, we have chosen a heptagon representation with seven variables, five of which are describing the values of scenario variables in the year 2100, one a cumulative figure up to that year, and one describing a growth rate.

The seven variables are (a) CO_2 emissions, expressed in billion (10^9) tons of carbon, (b) specific carbon emissions per unit of primary energy, expressed in grams of carbon per megajoule, (c) total primary energy consumption up to the year 2100, expressed in zetajoules (10^{21} joules), (d) specific primary-energy consumption per GDP, expressed in megajoules per US dollar (1990 purchasing power), (e) world gross domestic product, expressed in trillions (10^{12}) US dollars (1990 purchasing power), (f) population, expressed in billion (10^9) people, and (g) growth of world gross domestic product (GDP), expressed as the average annual growth rate (AAGR) of the GDP between 1990 and 2100. The resulting heptagon is presented in Figure 3.1.

The outer, regular heptagon (bold) represents the maxima for the group of all SRES database scenarios for each of the seven variables. The smallest, irregular heptagon inside this envelope represents the respective minima for all SRES database scenarios. Another irregular heptagon shows the respective seven median values for all SRES database scenarios. Similarly, for comparison, the remaining two heptagons show the seven minima and maxima for the group of sustainable-development (SD) scenarios for each of the seven variables.[2]

Care should be taken when interpreting the heptagons, in particular the one that connects the minima of the SRES database. It should be remembered that heptagons are unlikely to connect points that all belong to one and the same scenario. In fact, we would argue that it would not be logical

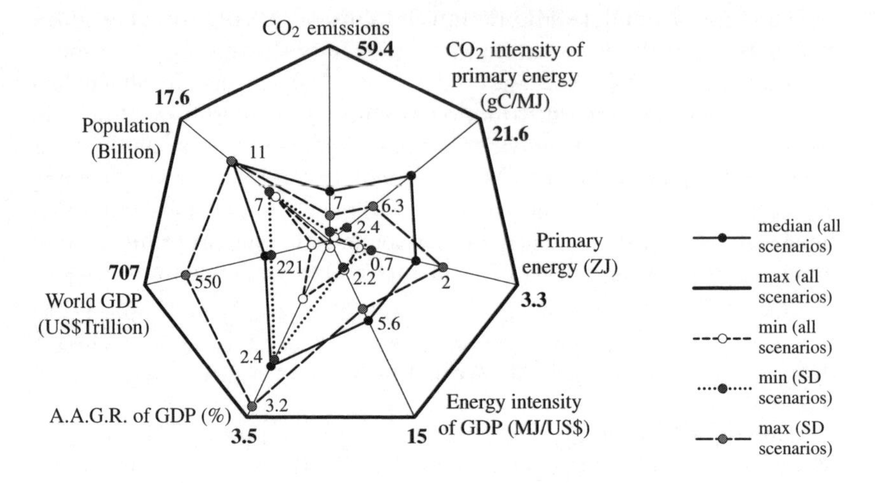

Note: The minimum and maximum of the values for the SRES database as well as for the SD scenarios are shown on seven axes of the heptagon, and they are connected. The seven axes show ranges for indicators across the scenarios in 2100.

Figure 3.1 Global carbon emissions and their main driving forces

if either a minimum or a maximum heptagon connected values for one and the same scenario. Take economic growth and primary-energy intensity, for example. One common assumption in E3 scenarios is that, with higher economic growth, technological progress can be faster. As a consequence, primary-energy intensity should not be maximum in scenarios with maximum economic growth.

Looking at the axes of Figure 3.1 gives an idea of the distributions of the seven indicators. These distributions can be quite asymmetric. CO_2 emissions in the year 2100, for instance, cover a range between zero (that is, no net carbon emissions) and almost 60 billion (10^9) tons of carbon (GtC), with the median at approximately 18 GtC, significantly below halfway. In contrast, world GDP growth and population both have a 'bias' towards the maximum. For population this is perhaps more obvious (global population is unlikely to be projected to reach any number near zero), but for GDP and GDP growth this means that the SRES database does not include scenarios with zero or low economic growth during the 21st century, which is still not surprising, but more noteworthy than for population.

The location of SD scenario values within the overall ranges is particularly interesting because it characterizes SD scenarios rather well. We therefore devote a full section (section 3.6 below) to the analysis of SD scenarios relative to all SRES scenarios and also in absolute terms.

3.2.1 Classifying Long-term E3 Scenarios

Sustainable-development scenarios

Sustainable-development scenarios are scenarios that fulfil the criteria in our definition (see page 5). They are based on a wide range of non-climate-policies[3] that aim at achieving sustainable development, most notably equity. The scenarios in this group are therefore all normative, that is, they describe desirable but not necessarily plausible developments of the global E3 system. These scenarios often lead to low GHG emissions levels.

CO_2 mitigation scenarios

The mitigation scenarios assume a constraint equivalent to climate policies that lead to a stabilization of atmospheric CO_2 concentrations. The most frequently used concentration limit in the model community at large and at IIASA is 550 ppmv. The reason for the popularity of this value is twofold. First, it roughly corresponds to twice the pre-industrial concentration level of 280 ppmv. This means that the common 'climate sensitivity' parameter – indicating the global temperature increase as a consequence of this doubling – directly corresponds to the value of 550 ppmv. Second, that level appears as a kind of first-order compromise between environmental and economic objectives.

The whole concept of considering concentration limits refers to the UN Framework Convention on Climate Change, which includes, as its central objective, the 'stabilization of greenhouse gas concentrations in the atmosphere at a level that would prevent dangerous anthropogenic interference with the climate system'. Since it is far from clear what level would achieve this goal, alternative concentration limits are used in carbon mitigation scenarios. For the IIASA scenarios, these were 450, 650 and 750 ppmv. In all mitigation scenarios, the atmospheric concentration of CO_2 is limited by a constraint on cumulative emissions, which is added to the baseline scenarios from the group of high-impact scenarios (described below).

Since climate policies will have significant consequences for sustainable development, the distinction between the scenarios of this group and sustainable-development scenarios is fine and to an extent arbitrary. We have removed doubts in favour of classifying none of the mitigation scenarios as a sustainable-development scenario. As illustrated below, this choice is justified by the fact that the mitigation scenarios typically include distinctly non-sustainable features. This observation supports our view that sustainable development is a more general goal than climate mitigation.

High-impact scenarios

This group comprises all scenarios that cannot be categorized in the first two groups. It is therefore the biggest group of scenarios in the SRES database and in the set of IIASA scenarios included in this book. The scenarios in this group include the so-called 'baseline reference scenarios', the name referring to typical purposes of the scenario's construction. Baselines are used for example to formulate the CO_2 mitigation scenarios described in the subgroup above. Reference scenarios are used in a more general sense to test the consequences of alternative assumptions, which often are normative. More as a result of this categorization, scenarios in this group turn out to be non-sustainable non-intervention scenarios. This group contains the scenarios with the highest GHG emissions.

3.3 IIASA'S LONG-TERM E3 SCENARIOS

One important purpose behind the compilation of the SRES database was to document the ranges of greenhouse gas emissions and their most important driving forces as they were published in the relevant literature. The SRES scenarios which were subsequently formulated were designed to cover most of the range of carbon dioxide, other GHGs, and sulphur emissions found in the SRES scenario database. Their spread is similar to that of the IS92 scenarios, which had also been prepared under the IPCC umbrella almost ten years earlier.

One might argue that leaving this range where it was almost ten years earlier may imply that modellers included assumptions that are currently considered rather improbable. We would not agree with this argument mainly because these ranges are quite large. Furthermore, note that there are many different possible combinations of emission driving forces ('scenarios') that produce the same emissions ranges. Finally, the overlapping emissions ranges of IS92 scenarios with those of most other global long-term scenarios in the literature may have to do more with modellers' behaviour than with the implied probabilities, which are essentially subjective.

Likewise, the IIASA scenarios presented in this book cover most of the ranges of carbon emissions and driving forces as they can be found in the SRES database. It is therefore not surprising that IIASA has developed scenarios in all three scenario groups defined in the previous section. We would also claim that, taken together, the IIASA scenarios are even representative of the three groups.

For the presentation in this book, we selected altogether 34 scenarios that have been developed and published by the ECS Program at IIASA since

1998. These include scenarios developed with the World Energy Council (WEC) in 1998 (Nakićenović *et al.*, 1998), scenarios developed for the IPCC Special Report on Emissions Scenarios (SRES) (Nakićenović and Swart, 2000), as well as scenarios developed for the impact assessment for Working Group III of the IPCC Third Assessment Report (TAR) (Metz *et al.*, 2001). Of the 34 scenarios, 13 were used for the cluster analysis described in the next chapter. The 34 scenarios can be classified into three subgroups.

- Seven sustainable-development scenarios, described in more detail below (section 3.6). In these sustainable-development scenarios, relatively low GHG emissions levels are achieved.
- Nineteen GHG mitigation scenarios. These scenarios explore cases in which the global atmospheric CO_2 concentration is stabilized at various levels.
- Eight high-impact (non-sustainable, non-intervention) scenarios.

Most of the 34 scenarios either belong to the SRES scenarios or are related to them (for example, mitigation scenarios based on SRES scenarios). At the beginning of the descriptions of the individual scenarios we therefore give an overview of the so-called 'four storylines' and the related four SRES scenario families.

Each of the four is based on a common specification of the main driving forces. Schematically, the four families can be depicted as branches of a two-dimensional 'tree' (Figure 3.2).

The two dimensions shown indicate the economic development–environmental (A–B) and the global–regional (1–2) orientation, respectively. The A1 family thus describes a world featuring high economic growth and a large degree of global collaboration whereas the B2 family describes a world with high consideration of environmental goals in a 'regionalized' world. The other two families show the 'mixed' cases of a regionalized world with high economic growth (A2) and an environmentally conscious world with a high degree of global collaboration (B1).

Closest to 'business-as-usual' (expressed in terms of medians of the SRES database) is B2. The highest number of energy supply options was assumed to be plausible in A1. The latter is therefore the scenario family with the most members, each describing a distinct energy supply strategy. The whole set of IIASA-ECS scenarios will now be described by taking one group of scenarios at a time.

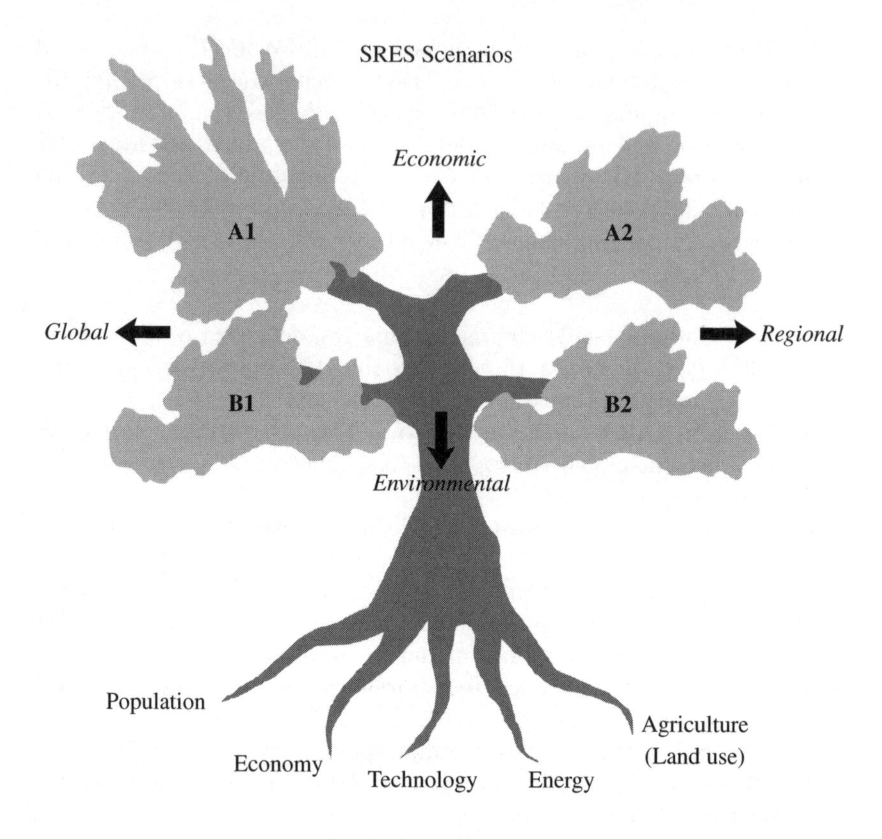

Note: The four scenario 'families' are arranged in two dimensions, one defined by an economic–environmental ('A–B') axis and the other by a global–regional ('1–2') axis.

Figure 3.2 Schematic illustration of SRES scenarios

3.3.1 Brief Characterization of High-impact Scenarios

The SRES A1 family contributes three scenarios to this group and one to the group of sustainable-development scenarios, depending on the energy supply system of each scenario. All four scenarios of this group are characterized by very rapid economic growth,[4] combined with low population growth (Lutz *et al.*, 1996, 1997). World regional average incomes per capita converge to the extent that 'poor' countries virtually disappear. This achievement is assumed to be based on a strong commitment to market-type solutions. In addition, the A1 world is characterized by a strong commitment to education, high rates of investment and increased international mobility of people, ideas and technology, accelerated by advances in communication technologies. In comparison to the other 'global economy' family (the more 'green' B1), the A1 world exhibits much higher energy demands, a consequence of assuming lower energy prices. Also the higher incomes assumed for A1 encourage comfortable and convenient (often energy-intensive) lifestyles.

A1B

The A1B scenario features 'balanced' progress of all primary-energy resources and all energy conversion technologies from energy supply to end use. Specific technology costs decrease significantly[5] for solar photovoltaic, fuel cells, hydrogen and wind technologies, and for liquid fuel production from coal or oil/gas. The A1 scenario assumes plentiful energy resources and high improvement rates for extraction, conversion and transport technologies. This initially results in the use of large quantities of hydrocarbon fuels, which later are increasingly replaced by options that do not emit carbon dioxide. Annual global CO_2 emissions peak at 20 GtC by 2060, and decline thereafter to 14 GtC in 2100 (see Figure 3.8).

A1C

The A1C scenario is dominated by 'clean coal' technologies. A1C illustrates the long-term implications of a rapid exhaustion of conventional oil and gas reserves, combined with slow progress in developing carbon-free alternatives, except for relatively high-cost improvements in new and clean coal technologies such as highly efficient high-temperature fuel cells, integrated coal gasification combined-cycle power plants (IGCC) and coal liquefaction. Methanol is becoming an important final-energy carrier, which is traded globally on a large scale. Not surprisingly, A1C is the scenario with the highest GHG emissions of all IIASA scenarios included in this book, with annual global CO_2 emissions approaching 33 GtC by 2100 (Figure 3.8).

A1G

The A1G scenario describes an oil and gas-rich future. In A1G, conventional resources of oil and gas are gradually but quickly replaced by abundant unconventional resources, including natural gas hydrates, oil shale and tar sands. This shift is driven by rapid technological progress in oil and gas extraction and conversion technologies. The extension of currently existing oil and gas grids and the construction of new natural gas pipelines from Siberia, the Caspian and the Middle East to China, Korea, Japan and South Asia (India) after 2020 lead to large-scale gas and oil trade with the natural gas share in global energy supply surpassing that of crude oil in the year 2030. CO_2 emissions approach the comparatively high level of about 31 GtC by 2100 (Figure 3.8).

A2

The A2 scenario foresees future developments towards a very heterogeneous (regionalized) world, characterized by high population growth in the developing regions, self-reliance in terms of resources, and less emphasis on economic, social and cultural interactions between world regions. Eventually, the world 'consolidates' into a series of economic trade blocks. Compared to the other scenarios in this high-impact group, A2 is characterized by relatively slow capital stock turnover, slower technological change and a more slowly narrowing income gap between today's industrialized and developing countries. High-income but resource-poor regions shift towards advanced post-fossil technologies, while low-income resource-rich regions generally rely on traditional fossil technologies. This leads to steadily increasing levels of GHG emissions (Figures 3.8 and 3.10), with CO_2 emissions approaching 28 GtC in 2100.

B2

The B2 world is one of high concern for environmental and social sustainability. In contrast to the sustainable development B1 scenario, however, international institutions decline in importance, with a shift towards local and regional decision-making structures and institutions, which favours local and regional pollution control. In the B2 world, most of the world's economic growth takes place in today's developing countries, leading to a moderate convergence in productivity and income levels over world regions.

In terms of population, technological change and energy use, B2 is clearly a 'dynamics-as-usual' scenario. Population follows historical trends (including recent faster-than-expected earlier fertility declines) towards a completion of the demographic transition within the next century (UN, 1998). This refers to a transition from high fertility and high mortality (resulting in low or no population growth) to high fertility and low mortality (resulting in rapid population growth as in most of today's developing countries or in Western Europe in the second half of the 19th century) and finally to low fertility and low mortality as in today's OECD countries (resulting in stable population size).

Between 1990 and 2100, global primary-energy demands increase by a factor of four, mainly owing to demand increases in today's developing regions. Cost reductions of most technologies are moderate (Table 3.1; SRES, Nakićenović and Swart, 2000). However, they are significant in particular for wind and solar photovoltaic, but also for gas combined-cycle, integrated gasification combined-cycle (IGCC), solar thermal power plants and advanced nuclear power plants.[6] Global GHG emissions in B2 increase approximately along a straight line (Figures 3.8 and 3.10), with CO_2 emissions reaching 14 GtC by 2100.

IIASA-WEC A

The IIASA-WEC A family of scenarios features high rates of economic growth, that is, an average annual rate of 2 per cent per year in the OECD countries and twice this rate in the developing countries. To achieve this, the scenario assumes a favourable geopolitical environment and free markets. The assumed high economic growth facilitates a more rapid turnover of capital stock and thus rapid technological progress. The IIASA-WEC A family of scenarios has three members, mainly distinguished by the dominant source of primary-energy supply. Two of the scenarios, A1 and A2, belong to the high-impact group. The third one, A3, is a sustainable-development scenario according to our classification.

IIASA-WEC A1

The IIASA-WEC A1 scenario assumes a high future availability of oil and gas resources, both conventional and unconventional. Oil and gas therefore dominate the global primary-energy supply up to the end of the 21st century.

IIASA-WEC A2

The IIASA-WEC A2 scenario is one in which greenhouse warming is little cause for concern, therefore leaving little incentive to phase out fossil fuels early, particularly in areas endowed with large, cheap coal resources. Sulphur and nitrogen emissions are mitigated through control technologies, and coal's vast resources make it the most preferred fossil fuel. Coal-based liquid fuels substitute for dwindling resources of conventional oil and gas, which are assumed to be limited to currently known reserves and resources.

IIASA-WEC B

The IIASA-WEC B scenario incorporates modest economic growth and modest technological development. The 'South' develops to some degree, but for some regions such as Africa, progress is rather slow. Together, these assumptions lead to relatively modest energy demand. In particular, slower technology improvements result in a high reliance on fossil fuels. Up to 2020, the structure of energy supply and end use remains closer to the current situation. After that time, oil and gas maintain a significant share in the global primary-energy mix up to about 2070. This is made possible because costlier categories of conventional and unconventional resources

Table 3.1 Technology improvement rates in IIASA's high-impact scenarios relative to all SRES scenarios

	Technology Improvement Rates			
	Coal	Oil	Gas	Non-fossil
SRES-A2	Average	Low	Low	Low–Average
SRES-B2	Low	Average	Average–High	Average
SRES-A1B	High	High	High	High
SRES-A1G	Low	Very High	Very High	Median
SRES-A1C	High	Low	Low	Low
WEC-A1	Average	High	High	Average–Low
WEC-A2	High	Low	Low	Low
WEC-B	Low–Average	Low	Average	Average

are being utilized. Constraints on the expansion of the production of fossil fuels prove to be based less on geology and more on financial and environmental considerations.

An overview of the assumptions on primary-energy technology development in the high-impact scenarios is given in Table 3.1. Selected drivers and results of the high-impact IIASA scenarios are summarized in Table 3.2.

3.3.2 Brief Characterization of CO_2 Mitigation Scenarios

Using the high-impact scenarios A2, B2, A1 and A1C[7] as baselines, IIASA-ECS developed a total of 17 CO_2 stabilization scenarios.[8] These scenarios were constrained to stabilize atmospheric CO_2 concentrations at levels of 450, 550, 650 and 750 ppmv in 2100. To limit the possible causes of mitigation costs to technology and fuel substitution we assumed the same set of technology data and the same set of resource availabilities as for the corresponding baseline scenarios.[9]

As MESSAGE variables do not include atmospheric CO_2 concentration, the constraint-limiting concentration was implemented in the model as a weighted sum of cumulative CO_2 emissions from 1990 to 2100. This approximates the effects of the carbon cycle by using a time profile of CO_2 absorption by a variety of sinks, most notably the oceans. This approach differs from approaches by other modellers who use given emission trajectories known to stabilize concentrations at a given level, such as the popular WRE trajectories (Wigley *et al.*, 1996), as constraints on annual emissions. Using trajectories constrains carbon emissions in each time period and thus eliminates the flexibility in time of emission mitigation, but appears to us as a reasonable simplification. As we shall illustrate below, our emission paths are very similar to the equivalent WRE paths.

Selected drivers and results of the IIASA CO_2 mitigation scenarios with a constraint on atmospheric CO_2 concentrations at levels of 550 ppmv in 2100 are summarized in Table 3.3. Constraining emissions in MESSAGE without at the same time allowing for the deployment of additional technologies results in higher energy supply costs, which in turn are expected to lead to lower demands. For this reason, the stabilization scenarios were generated using the MESSAGE-MACRO model, which is briefly described in the appendix to this book. The model results are therefore cost-optimal actions to meet the given carbon constraint and in consideration of a price responsiveness of energy demand.

In addition to the temporal flexibility of only constraint on cumulative emission, MESSAGE-MACRO also allows for spatial flexibility of emission reductions by and the free movement of investments across world regions. Cost-optimal CO_2 emission reduction therefore does not necessarily occur in regions that give high priority to such reductions and that have the money to pay for them. Rather, cheapest CO_2 reductions are implemented first. The stabilization scenarios can thus be seen as possible answers to the question, 'Which are the best strategies to achieve stabilization *if* the world, generally consistent with the (respective) baseline, was able successfully to coordinate and cooperate on efforts to limit potential global warming?'[10]

For a more detailed description of the IIASA mitigation scenarios see Roehrl and Riahi (2000).

3.3.3 Brief Characterization of Sustainable-development Scenarios

All scenarios of the B1 family belong to the group of sustainable-development scenarios. This result is achieved by assuming service-oriented prosperity, while taking into account equity and environmental concerns without policies directed at mitigating climate change. Telecommunications and information technology expand rapidly, giving less developed regions important opportunities to progress. Economic production is thus characterized by rapid 'dematerialization' and the introduction of clean technologies eventually leads to hydrogen-based economies in all world regions.

B1 describes a rapidly converging world, emphasizing global solutions for environmental and social sustainability, including concerted efforts aiming at rapid technology development, technology transfer, dematerialization of the economy, and improving equity (both worldwide and within regions). As in the high-impact A1 scenarios described above, world

Table 3.2 Selected drivers and results of the high-impact scenarios

Scenario	Population, billion (10^9)			Gross domestic product[a] (GDP) US$(1990) trillion ($10^{12}$)			Equity[b]		Primary energy demand (EJ)			Cumulative CO_2 emissions (GtC)	Atmospheric CO_2 concentration (ppmv)		SO_2 emissions (MtS)	Global temperature change[c] 1990 to 2100 (K)		
	1990	2050	2100	1990	2050	2100	1990	2100	1990	2050	2100	1990–2100	1990	2100	1990	2050	2100	2100
SRES-A2	5.3	11.3	15.1	20.9	82	243	0.06	0.24	352	1014	1921	1662	354	783	69	100	66	2.7
SRES-B2	5.3	9.4	10.4	20.9	110	235	0.06	0.33	352	869	1357	1143	354	603	69	54	45	2.0
SRES-A1B	5.3	8.7	7.1	20.9	187	550	0.06	0.64	352	1422	2681	1562	354	724	69	55	29	2.4
SRES-A1C	5.3	8.7	7.1	20.9	187	550	0.06	0.64	352	1377	2325	2046	354	950	69	122	47	3.0
SRES-A1G	5.3	8.7	7.1	20.9	187	550	0.06	0.64	352	1495	2737	2092	354	891	69	68	38	2.8
WEC A1	5.3	10.1	11.7	20.9	100	300	0.06	0.21	352	1048	1895	1441	354	620	69	54	23	2.3
WEC A2	5.3	10.1	11.7	20.9	100	300	0.06	0.21	352	1048	1896	1632	354	730	69	64	55	2.6
WEC B	5.3	10.1	11.7	20.9	73	202	0.06	0.21	352	837	1464	1138	354	590	69	55	58	2.1

Key: EJ = Exajoules; GtC = Gigatons of carbon; K = Degrees Kelvin; MtS = Million tons of sulphur; ppmv = parts per million by volume.

Notes:

[a] When not mentioned explicitly otherwise, gross domestic product (GDP) and related parameters are reported assuming market exchange rates.

[b] Ratio of per capita incomes between developing and industrialized regions.

[c] Assuming a climate sensitivity of 2.5 K (Wigley and Raper, 1997); an older version of the same model (Wigley *et al.*, 1994) was used for the WEC scenarios.

Table 3.3 Selected drivers and results of the CO_2 mitigation IIASA scenarios for stabilization at atmospheric concentrations of 550ppmv[11]

Scenario	Population, billion (10^9)			Gross domestic product (GDP) US$(1990) trillion ($10^{12}$)			Equity[a]		Primary energy demand (EJ)			Cumulative CO_2 emissions (GtC)	Atmospheric CO_2 concentration (ppmv)		SO_2 emissions (MtS)	Global temperature change[b] 1990 to 2100 (K)		
	1990	2050	2100	1990	2050	2100	1990	2100	1990	2050	2100	1990–2100	1990	2100	1990	2050	2100	2100
A2-550	5.3	11.3	15.1	20.9	81	236	0.06	0.23	352	959	1571	1210	354	550 (S)	69	81	54	2.1
B2-550	5.3	9.4	10.4	20.9	103	231	0.06	0.33	352	881	1227	971	354	550 (S)	69	56	38	1.8
A1B-550	5.3	8.7	7.1	20.9	186	547	0.06	0.63	352	1339	2505	1095	354	550 (S)	69	47	19	1.9
A1C-550	5.3	8.7	7.1	20.9	185	542	0.06	0.64	352	1269	2188	1093	354	550 (S)	69	71	30	2.0

Key: EJ = Exajoules; GtC = Gigatons of carbon; K = Degrees Kelvin; MtS = Million tons of sulphur; ppmv = parts per million by volume.

Notes:

[a] Ratio of per capita incomes between developing and industrialized regions.

[b] Assuming a climate sensitivity of 2.5 K (Wigley and Raper, 1997); with the same assumption, temperature change from 1765 to 1990 was 0.4 K.

population projections are low in all B1 cases. Global GDP is considerably higher in the B1 scenario family (that emphasises global solutions) than in the corresponding 'regionalized' scenario family B2. This is similar to the case of the A1 family, where global GDP is considerably larger than in the 'regionalized' scenario family A2.

Subsidies for traditional energy technologies and fuels are phased out, and capital markets increasingly respond negatively to environmental accidents. This leads to careful land management and the deployment of 'clean' energy technologies (Table 3.1 and Figure 3.9). The particular institutional developments assumed in the B1 world favour decentralized energy supply. The transport, residential/commercial and industrial sectors rely increasingly on fuel cell-based electricity generation. Resulting emission levels are among the lowest of all the scenarios considered here (Figures 3.8, 3.9 and 3.10). In particular, annual global CO_2 emissions range from 3 to 6 GtC in the year 2100.

Just like the A1 family, B1 includes several distinctly different energy supply scenarios, except that, in B1, all of them belong in the sustainable-development group. (There is only one SD scenario in the A1 family.) We now describe them in turn.

B1G

The B1G scenario explores a 'natural-gas and non-fossil future', in particular natural gas-based infrastructures as a transition to hydrogen as the eventually dominant fuel.

B1T

B1T illustrates a particularly rapid shift to non-fossil and decentralized technologies. B1T is a very optimistic case in which the world energy supply system 'leaps' directly to a hydrogen-based economy.

B1B

The B1B scenario features 'balanced' progress across all resources and technologies. In a sense, it is a blend of B1G and B1T.

A1T

A1T portrays a 'post-fossil' sustainable-development future with rapid cost decreases of solar and advanced nuclear technologies[12] on the supply side, and mini-gas turbines and fuel cells used in energy end use applications. In contrast to the B1 scenarios, A1T is characterized by very rapid economic growth and, hence, also comparatively high energy demands (as is the case also in the other A1 variants). A1T assumes medium levels of availability of oil and gas. This, together with the relatively fast turnover of capital, leads to the rapid diffusion of carbon-free and advanced decentralized technologies (for example, solar PV), particularly in the second half of the century (see Figure 3.3). Resulting CO_2 emissions peak at 13 GtC in 2050, and decline thereafter to about 5 GtC in 2100 (Figure 3.8). This scenario is described as the 'Post-Fossil' scenario in Chapter 5.

IIASA-WEC A3
As a member of the IIASA-WEC A family, the IIASA-WEC A3 scenario features high rates of economic growth and rapid technological progress, in particular of nuclear and renewable energy technologies results. Accordingly, fossil fuels are phased out for economic reasons rather than because of resource scarcity.

IIASA-WEC C
The IIASA-WEC C family of scenarios is optimistic about technology and geopolitics, assuming unprecedented progressive international cooperation focused explicitly on environmental protection and international equity. Among others, it includes 'green' taxes, substantial resource transfers from industrialized to developing countries, spurring growth in the South. IIASA-WEC C incorporates policies to reduce carbon emissions in 2100 to 2 GtC per year. The two scenarios of this family reflect two possible developments of nuclear energy technology.

IIASA-WEC C1
The IIASA-WEC C1 scenario reflects the present reservations of environmentalists against nuclear energy. It assumes that the public acceptance of this technology will remain low and that therefore nuclear energy is phased out entirely by the end of the 21st century.

IIASA-WEC C2
In IIASA-WEC C2, a new generation of advanced nuclear reactors is developed. The basic role of nuclear energy is the same as in A1T (see the description above), that is, it is widely accepted.

Selected drivers and results of the sustainable-development IIASA scenarios are summarized in Table 3.4. An overview of the assumptions on primary-energy technology development in the sustainable-development scenarios is given in Table 3.5.

3.4 RESULTS OF ALL THREE SCENARIO SETS

The overall supply characteristics of an E3 scenario are best illustrated in a so-called 'energy triangle'. Energy triangles are an example of graphically

Table 3.4 Selected drivers and results of the sustainable-development IIASA scenarios

Scenario	Population, billion (10^9)			Gross domestic product (GDP) US$(1990) trillion ($10^{12}$)			Equity[a]		Primary energy demand (EJ)		
	1990	2050	2100	1990	2050	2100	1990	2100	1990	2050	2100
SRES-A1T	5.3	8.7	7.1	20.9	187	550	0.06	0.64	352	1213	2021
SRES-B1	5.3	8.7	7.1	20.9	136	328	0.06	0.59	352	837	755
SRES-B1G	5.3	8.7	7.1	20.9	166	350	0.06	0.60	352	911	1157
SRES-B1T	5.3	8.7	7.1	20.9	136	328	0.06	0.59	352	819	714
WEC-A3[d]	5.3	10.1	11.7	20.9	100	300	0.06	0.21	352	1040	1859
WEC-C1	5.3	10.1	11.7	20.9	75	220	0.06	0.35	352	601	881
WEC-C2	5.3	10.1	11.7	20.9	75	220	0.06	0.35	352	601	880

Key: EJ = Exajoules; GtC = Gigatons of carbon; K = Degrees Kelvin; MtS = Million tons of sulphur; ppmv = parts per million by volume.

Notes:
[a] Ratio of percapita incomes between developing and industrialized regions.
[b] Sulphur emissions for the WEC scenarios include energy-related emissions only
[c] Assuming a climate sensitivity of 2.5 K (Wigley and Raper, 1997); with the same assumption, temperature change from 1765 to 1990 was 0.4 K.
[d] Note that the WEC A3 scenario has been classified as a sustainable development scenario mainly because of its environmental sustainability. Also the socioeconomic gap between North and South is closed considerably, although to a lesser extent than in the other sustainable development scenarios.
'S' denotes those scenarios where CO_2 concentrations are stabilized in 2100.

Table 3.5 Technology improvement rates in IIASA's sustainable-development scenarios relative to the range of all SRES scenarios

	Technology Improvement Rates			
	Coal	Oil	Gas	Non-fossil
SRES-A1T	Low	High	High	Very High
SRES-B1	Low–Average	Average–High	High	High
SRES-B1G	Low–Average	Average–High	High	High
SRES-B1T	Low–Average	High	High	Very High
WEC-A3	Low–Average	Low	High	Average–High
WEC-C1	Low–Average	Low	Average–High	High
WEC-C2	Low–Average	Low	Average–High	High

Table 3.4 (continued)

Cumulative CO_2 emissions (GtC)	Atmospheric CO_2 concentration (ppmv)		SO_2 emissions[b] (MtS)	Global temperature change[c] 1990 to 2100 (K)		
1990–2100	1990	2100	1990	2050	2100	2100
1122	354	560	69	41	17	1.9
842	354	486 (S)	69	28	9	1.7
902	354	509	69	31	13	1.8
776	354	464 (S)	69	27	8	1.6
1072	354	550 (S)	69	45	9	2.1
635	354	445 (S)	69	22	7	1.5
622	354	445 (S)	69	22	5	1.5

illustrating three variables, which cannot freely take arbitrary values because they are all positive and required to add up to 100 per cent. Because of this restriction, they can be plotted in two dimensions without loss of information. Our energy triangle plots shares of three kinds of primary energy: coal, oil plus gas, and non-fossil primary energy (Figure 3.3).

For the years 1990 to 2100, quite distinct trajectories unfold for the five baseline reference scenarios A2, B2, A1, A1G and A1C (long dashes) and for the four 550 ppmv CO_2 concentration stabilization cases A2-550, B2-550, A1-550 and A1C-550 (short dashes). Only the four sustainable-development scenarios B1, B1G, B1T and A1T (dotted lines) show similar patterns of change. They first shift rapidly toward gas and later toward zero-carbon options.

The interpretation of Figure 3.3 is best approached by focusing on the three vertices of the triangle. Each of them represents a situation in which one of the three kinds of primary energy has a share of 100 per cent (with no contributions from the other two). In the inner area of the triangle, the share of each kind of primary energy is reflected by the distance of a point from the line opposite the vertex corresponding to that particular kind of primary energy. To facilitate reading the graph, we have plotted iso-lines at distances of 20 per cent.

Figure 3.3 shows primary-energy supply over time as curves within the energy triangle. Since the triangle itself does not have a time dimension, selected points in time are marked on the curves. The graph shows one actual development (for the time between 1850 and 1990) and time paths for 13

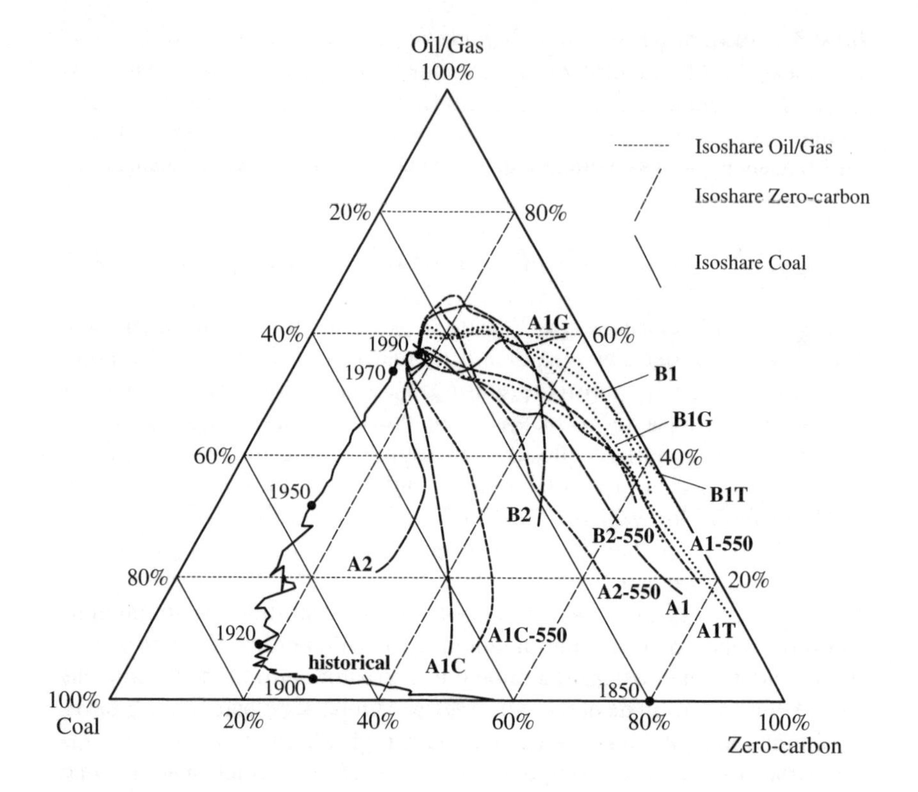

Note: Historical data from 1850 to 1990 are based on Nakićenović *et al.* (1998).

Figure 3.3 Shares in primary energy

representative scenarios. The curve depicting the actual past development begins in 1850 with a point representing 80 per cent of global energy being supplied by non-fossil energy (fuel wood in this case) and 20 per cent by coal. Between 1850 and approximately 1910, coal substitutes for fuel wood, reaching a peak of approximately 75 per cent. Non-fossil energy still has a 20 per cent share, and oil plus gas covers the remaining 5 per cent.

The time period between 1910 and 1970 may be called the 'oil era'. In those 60 years, oil plus gas reach a share of approximately 55 per cent while coal's share drops to 30 per cent. Non-fossil energy by and large keeps its 15 per cent share. The three shares do not change much between 1970 and 1990 – at least much less than during the 20 years before that. For the period after 1990, Figure 3.3 illustrates the wide range of energy supply patterns in the IIASA scenarios. The two most extreme scenarios are A2 on the 'coal' side and A1T on the 'non-fossil' side, the latter being the sustainable development scenario that is described in detail in Chapter 5. In the remainder of this chapter, we continue the overview description of all IIASA scenarios together.

3.4.1 Medians and Ranges of Market Shares of Energy Technologies

The purpose of determining market shares of energy technologies is to establish a basis for the design of technological strategies. For the analysis, we use so-called 'technology clusters', that is, representative aggregate technologies.[13] For each scenario set considered here, we determine the minimum, maximum and median of future market shares for each technology. The frequent occurrence of a technology in any given scenario set is interpreted as a high future potential of the technology in the particular 'world' as defined by that set, and a technology that contributes substantially in all scenario sets is considered a robust future technology option. Smaller ranges[14] around the median market share of a technology enable us to have higher confidence in the size of the future market share than larger ranges.

The analysis of technology shares and frequencies also determines the relative importance of the electricity generation technologies considered in this analysis. Assessing the importance of the technologies considered here, it should be kept in mind that there remains the possible availability of improved or completely new, or not yet even conceived, technologies that are not included explicitly in our scenario sets. However, this is not considered a serious shortcoming, for two reasons. First, the uncertainty concerning the future availability of new technologies is not a great problem because world market diffusion rates for such new technologies are extremely low. As a 'rule of thumb', it takes about 50 years for a technology to proceed from a 1 per cent to a 10 per cent market share (Marchetti and Nakićenović, 1979; Marchetti, 1980).

Second, the interest here is mainly in identifying robust *patterns* in the dynamics. For this purpose, an understanding of the evolvement of technology clusters appears more important than the absolute values of market shares of single technologies. In other words, only completely new technologies that would not be part of any existing technology cluster could produce a dynamics significantly different from our model, whereas the dynamics induced by new, not yet conceived but based on existing technology clusters would still be covered by the model in an order-of-magnitude fashion.

For each scenario set, the medians and ranges of future market shares for each aggregated technology are shown in Figures 3.4, 3.5 and 3.6 for the years 2030, 2050 and 2100, respectively. The abbreviations describing the aggregate technologies (on the horizontal axis) are explained in Table 4.1. The future market shares displayed in these three figures should be seen in

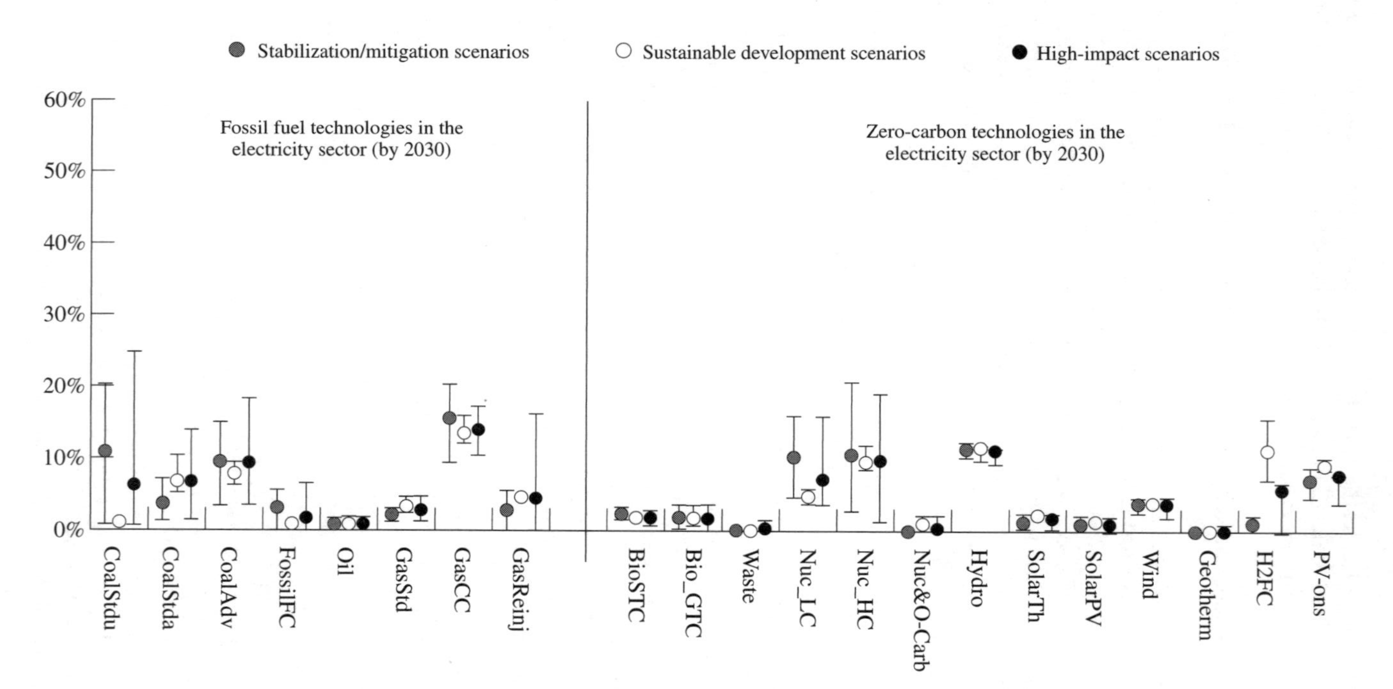

Figure 3.4 *Medians and ranges of market shares (in percentage of global electricity production) of aggregate technologies in three scenario sets in 2030*

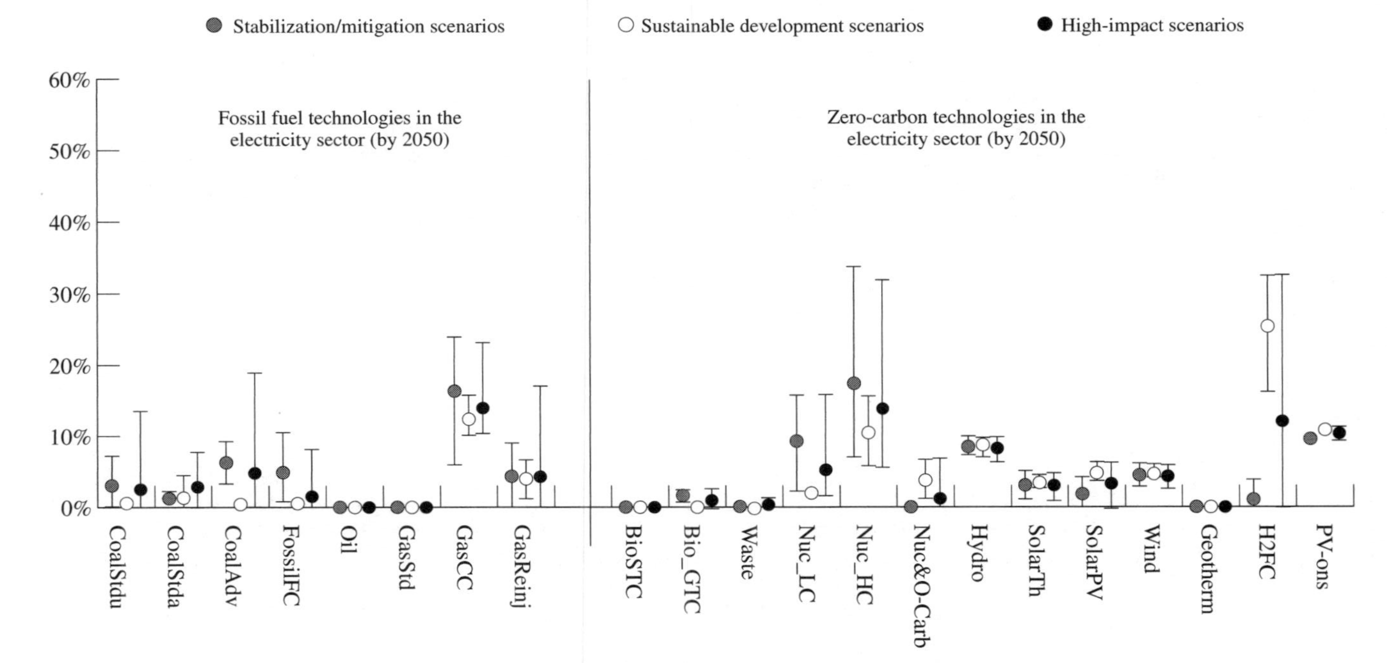

Figure 3.5 Medians and ranges of market shares (in percentage of global electricity production) of aggregate technologies in three scenario sets in 2050

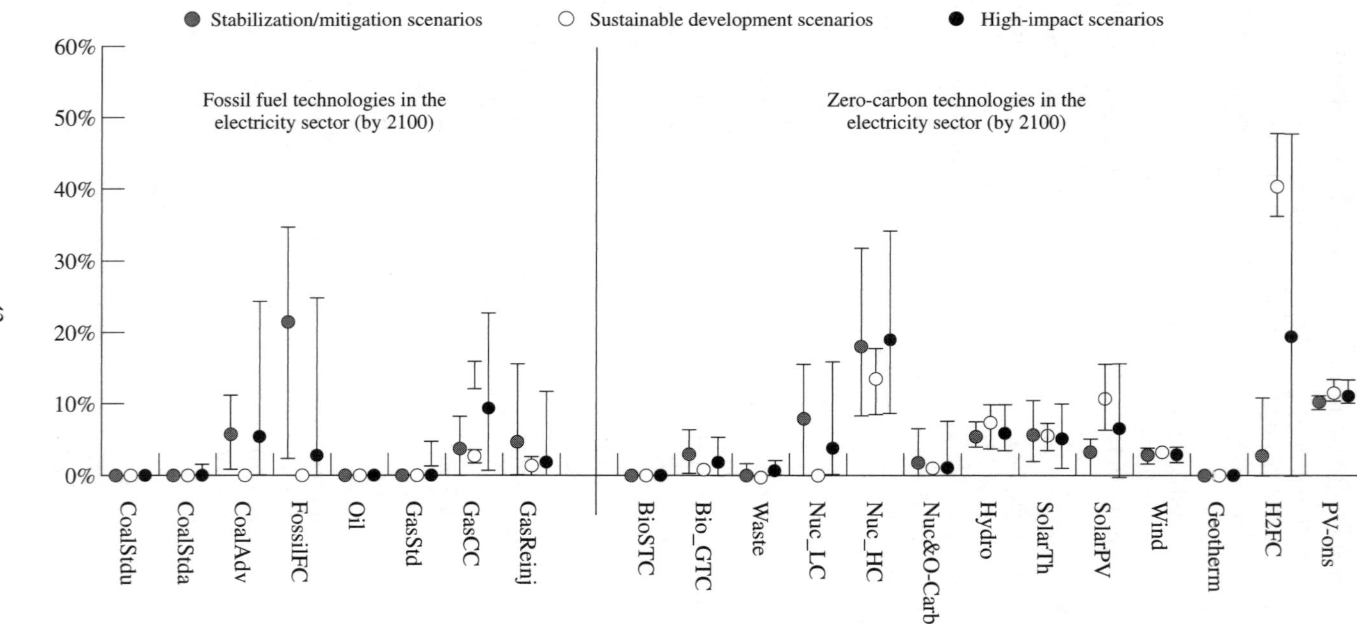

Figure 3.6 Medians and ranges of market shares (in percentage of global electricity production) of aggregate technologies in three scenario sets in 2100

relation to the structure of global electricity generation in 1990, which was dominated by fossil fuels (65 per cent of total electricity output) and where nuclear (17 per cent) and hydropower (18 per cent) supplied most of the balance. This comparison shows that, in almost all scenarios analysed here, the technological structure of the power sector changes significantly during the 21st century.

Naturally, the emerging structures are different in the high-impact scenarios, the sustainable-development scenarios and the CO_2 mitigation scenarios. Of these three sets, the high-impact scenarios show the widest ranges of market shares for almost all technologies, indicating the high overall uncertainty surrounding the adoption of some electricity technologies in scenarios that are classified essentially according to their lack of policies with regard to climate mitigation.

There are two notable exceptions to this general observation. The ranges for gas combined-cycle power plants (GasCC) in 2050, and fuel cells based on fossil fuels (FossilFC) in 2100 are larger in the CO_2 mitigation scenarios than in the high-impact scenarios. The reason for the first exception is that, in some mitigation cases, gas combined-cycle power plants are used to replace less efficient and more carbon-intensive electricity generation from coal. Therefore it is possible that the maximum contribution in the mitigation scenarios is even larger than that for the high-impact scenarios. The minimum is lower than for the high-impact scenarios because in some mitigation scenarios (those that require a very rapid introduction of non-fossil technologies) the replacement of GasCC begins before 2050. The reason for the wide range of fossil fuel-based fuel cells is that, for some mitigation scenarios, highly efficient fuel cells in combination with carbon scrubbing play an important role. Hence the maximum and minimum contributions of fossil fuel cells exceed the range of their contributions in the high-impact scenarios.

Figure 3.6 shows that, in all three scenario sets, conventional fossil-fuel power plants (conventional coal-fired, oil-fired and gas-fired power plants with a steam cycle) are phased out during the 21st century. They are gradually replaced by gas combined-cycle (GasCC) technology, which later gives way to more advanced fossil and non-fossil technologies. In many high-impact and in some mitigation scenarios, advanced fossil-based technologies become an important option by 2100. In the mitigation scenarios, this option generally includes carbon scrubbing. Accordingly, the ranges for fossil-based fuel cells, for gas combined-cycle and for advanced coal technologies (for example, IGCC) are particularly wide in these two scenario sets.

The sustainable-development (SD) scenarios feature much more narrow ranges for the future market shares of fossil-fuel power plants than the other two scenario sets. In the SD scenarios, the only relevant fossil fuel in 2100 is natural gas, and its market share in 2100 is rather small compared to non-fossil options (Figure 3.6). By the same token, the SD scenarios feature narrow ranges around high median future market shares of hydrogen-based fuel cells. Their minimum share in total electricity generation (in this set of scenarios) increases from 18 per cent in 2050 to 35 per cent in 2100.

The mitigation scenarios show rather small market shares (up to 11 per cent) of fossil-fuel power plants, and their market shares spread over a wide range in high-impact scenarios (ranging from zero per cent to 49 per cent in 2100). Note that hydrogen fuel cells do not emit any carbon at the level of electricity production. However, the production of hydrogen can cause carbon emissions, for instance, when hydrogen is produced from fossil fuels. Consequently, hydrogen fuel cells may only be regarded as a non-fossil electricity generation option when carbon-free fuels are used also for the production of hydrogen. (Figure 3.7 provides an overview of the sources of hydrogen production in the scenarios.)

Today, nuclear power faces significant public opposition, mostly as a consequence of concerns about the safety of the technology including the nuclear fuel cycle. In the IIASA scenarios, it is therefore assumed that in the future new nuclear technology, for instance, inherently safe reactors will be available. This technology ('Nuc_HC') makes significant contributions in all three scenario sets. The minimum share of this technology over all three sets is approximately 9 per cent in 2100 (see Figure 3.6). Its maximum contribution is 35 per cent (high-impact and mitigation scenarios). The maximum share of Nuc_HC in sustainable-development scenarios is significantly less, i.e., below 20 per cent. This reflects the doubt, on the side of sustainable-development proponents, that the utilization of nuclear energy is sustainable.

A robust conclusion from this analysis of market shares is that hydrogen fuel cells are the only dominant technology in the technology menu considered. Still, there is no scenario in which only one or two zero-carbon options dominate. All scenarios feature a mixture of more than two zero-carbon options. For the world as a whole, the total contribution from carbon-free power (hydropower, wind, solar and biomass technologies) increases production substantially in all scenarios. In different world regions, the economic and technical potentials for the zero-carbon options such as hydropower, wind, solar and biomass can differ quite substantially (for example, the solar-energy potential in the Sahara is much larger than in more northerly regions).

We argue that, overall, the scenario results suggest a 'robust' mix of future carbon-free technologies in the electricity sector. The robustness is the highest in sustainable-development scenarios, followed by the CO_2 mitigation scenarios and the high-impact scenarios. Since hydrogen plays such a dominant role, we now look in more detail at the hydrogen production in the IIASA scenarios.

3.4.2 Sources of Hydrogen

Figure 3.7 shows the annual amounts of hydrogen production from coal, natural gas, solar, and other sources, respectively. In general, the share of carbon-free hydrogen production increases with time. As to primary-energy sources, coal plays only a minor long-term role in the hydrogen production in all scenarios. In many scenarios, steam reforming of natural gas plays an important short- to medium-term role in the transition to zero-carbon sources of hydrogen such as solar, which turns out to be the main source of hydrogen production in the second half of the 21st century, in particular in the sustainable-development scenarios, the A1B variants, as well as in A2-550 and B2-550.

3.4.3 Environmental Impact

CO_2 emissions in scenarios of all three groups are illustrated in Figure 3.8. Note that the sustainable-development scenarios show emissions trajectories similar to the CO_2 mitigation scenarios. Both scenario sets, the sustainable-development and the stabilization scenarios, cluster in the range of 4 to 7 GtC in 2100.

Current annual anthropogenic sulphur emissions have been estimated at between 65 and 85 million tons (MtS), for instance by Smith *et al.* (1998) and Grübler (1998). In comparison, natural emissions have been estimated to range between 4 and 45 MtS (Houghton *et al.*, 1995). Concerning future emissions of sulphur, our scenarios estimate global anthropogenic emissions of 30 to 120 MtS by 2050 and between 9 and 65 MtS by 2100 (see Figure 3.9).

Sooner or later, sulphur emissions begin to decrease in all scenarios, ranging from immediate decrease in the B1 scenarios to more gradual, later and less stringent controls in the A2 scenario, for instance. This pattern reflects both the impact of recent legislation for a drastic reduction in sulphur emissions in OECD countries as well as an anticipated gradual introduction of sulphur control policies in developing regions in the long term.

It is important to note that all these scenarios are sulphur-control scenarios only and do not assume any additional climate-policy measures. There is, however, a certain indirect GHG emission-reduction effect from sulphur-control policies leading to energy conservation and inter-fuel substitution from high-sulphur to low-sulphur fuels (for example, from coal to gas).

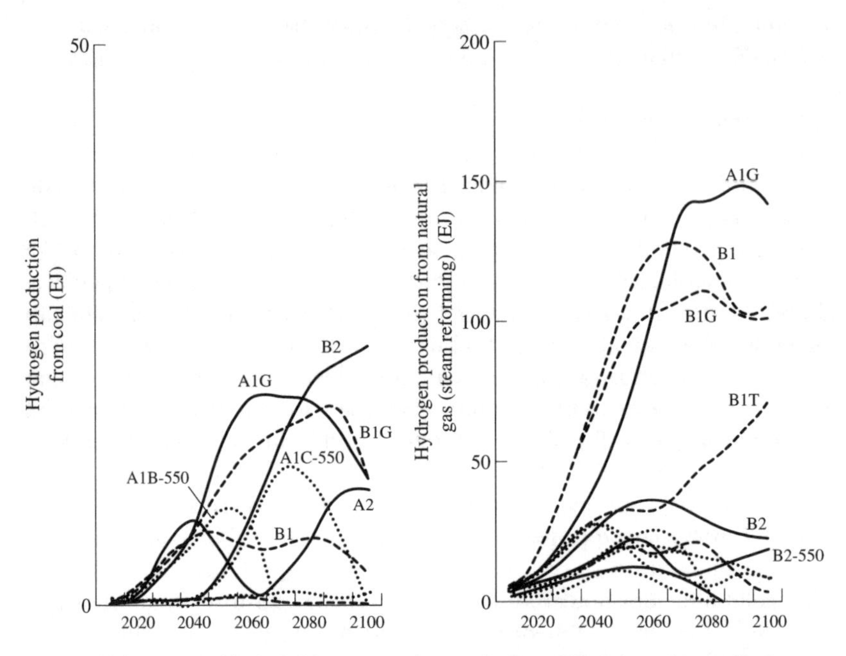

Note: 'Other sources' include biomass, nuclear, and others; EJ of secondary or final energy (depending on how the hydrogen is used, i.e., for electric and non-electric purposes).

Figure 3.7 Total annual hydrogen production from coal, natural gas (steam reforming), solar and other sources

The projected anthropogenic methane (CH_4) emission trajectories for the scenarios are displayed in Figure 3.10. Methane emissions in the year 1990 have been estimated at 375±75 $MtCH_4$ (SRES; Nakićenović and Swart, 2000). Methane emissions arise from a variety of human activities and, predominantly, biological processes, each associated with considerable uncertainty (Kram *et al.*, 2000). Our scenarios use the value of 310 Mt for the year 1990, which is within the range just mentioned. Approximately one-quarter of this is related to fossil-fuel extraction (methane emissions from coal mines, methane venting from oil extraction), transport and distribution (leakage from pipelines) and consumption (incomplete combustion). The biogenic sources of methane emissions include agriculture (enteric fermentation, rice paddies and animal waste), biomass burning, and waste from human settlements (landfills, sewage).

Hence the future trajectories of methane emissions depend in part on the volumes of fossil fuels used in the scenarios, adjusted for assumed changes in operational practices, but more strongly on scenario-specific, regional

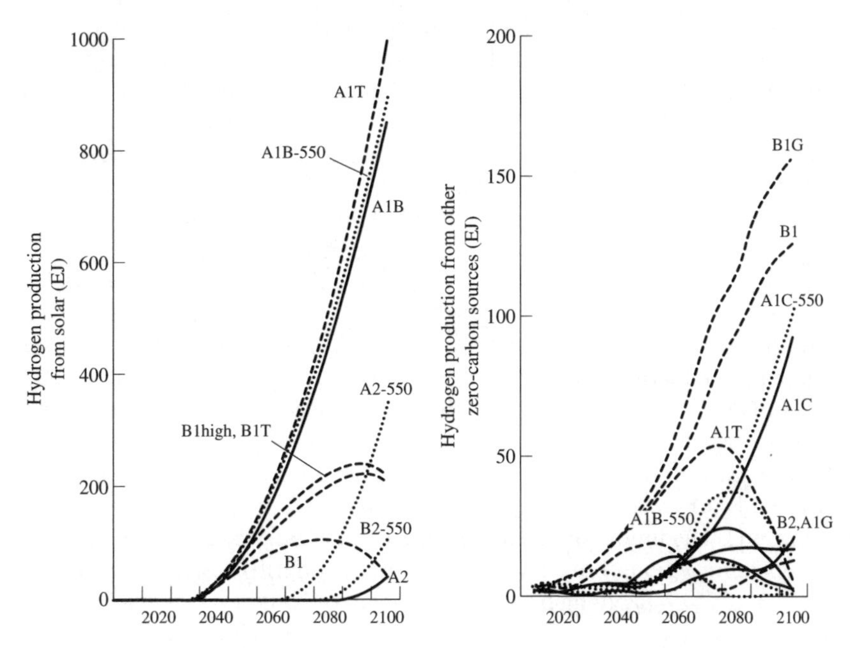

Figure 3.7 (*continued*)

demographic and affluence developments, together with assumptions on preferred diets and agricultural practices. As shown in Figure 3.10, the future development of methane emissions is associated with considerable uncertainties. Methane emissions in the high-impact scenarios range between 350 and 1070 Mt in the year 2100, and are well above the range for the sustainable development scenarios (250–300 Mt). As also illustrated in Figure 3.10, the stabilization of CO_2 emissions leads to ancillary benefits for CH_4 emission reduction (predominantly due to fuel switches in the energy sector). Limiting the CO_2 concentrations to 550 ppmv, for instance, leads to reductions of CH_4 emissions levels of between 50 and 300 Mt.[15]

Figure 3.11 shows global mean temperature changes relative to 1990 for all scenarios. These estimates are for a 'best guess' climate sensitivity of 2.5°C (Houghton *et al.*, 1996). However, this climate sensitivity parameter is highly uncertain. For this reason, it has often been suggested that a lower bound for climate sensitivity parameters of 1.5°C and an upper bound of 4.5°C (IPCC, 1996) are used. Using this range instead of the central value would embed our results in a rather wide (and perhaps confusing) 'uncertainty range' for global mean temperature change in 2100 (relative to 1990). For the B2 scenario alone, this range would reach from 1.4°C to 2.9°C (around a 'best guess' estimate of 2.0°C). Note that this uncertainty range

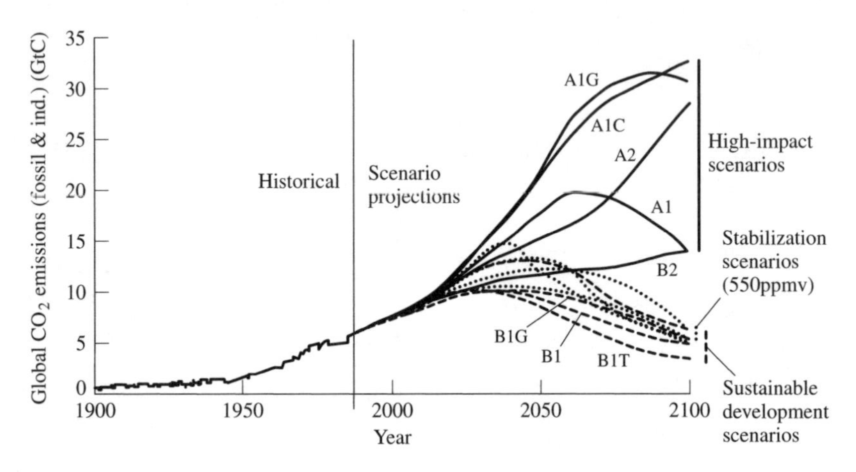

Note: Actual data from 1850 to 1990 are according to Marland *et al.* (1999).

Figure 3.8 Global CO_2 emissions from fossil fuel combustion and cement production for the three groups of future development scenarios (1990 to 2100)

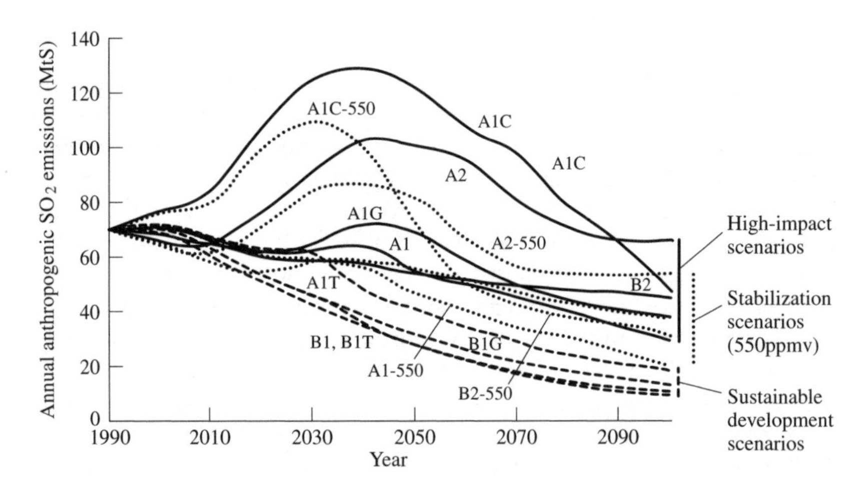

Figure 3.9 Projections of global, anthropogenic sulphur dioxide emissions

for B2 is practically identical to the range of best guesses over *all* scenarios (1.6°C for B1T to 3.0°C for A1C in 2100) discussed in this chapter.

Talking now only in terms of best-guess values, global mean temperature projections for 2100 show a range of increases from 1.6 to 1.9°C for the

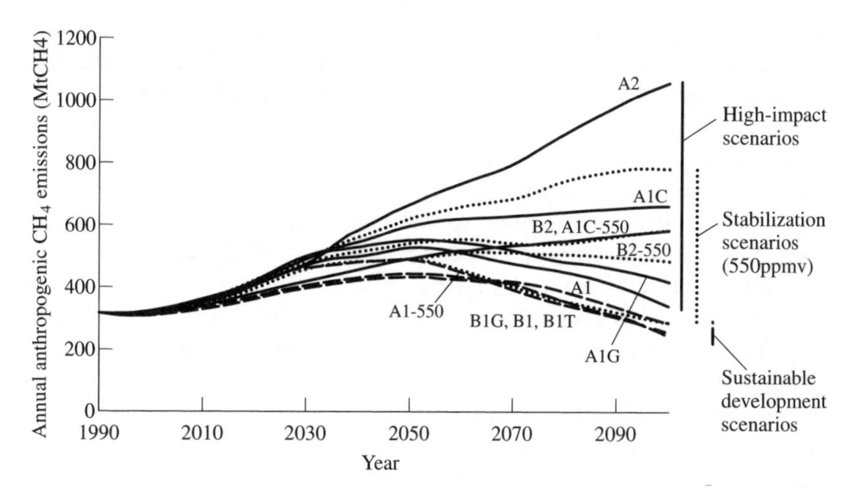

Figure 3.10 *Global anthropogenic methane emissions*

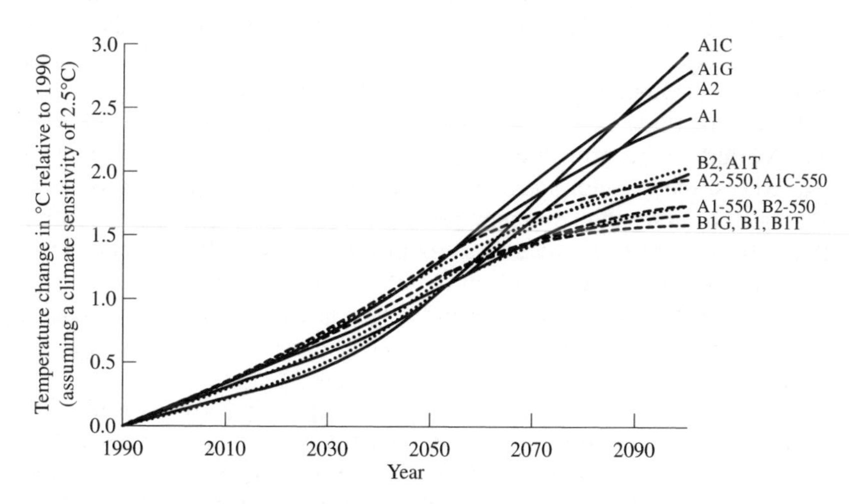

Figure 3.11 *Global mean temperature change in °C assuming an intermediate 'best guess' climate sensitivity of 2.5°C*

sustainable-development scenarios, from 2.0 to 3.0°C for the high-impact scenarios and from 1.8 to 2.1°C for those CO_2 mitigation scenarios that stabilize at 550 ppmv in 2100.

Before 2050, global mean temperature change is not very different for all the scenarios (Figure 3.11). This is due to the combined inertia of the energy system and the climate system, and to a balancing effect of sulphur emissions. Decreasing sulphur emissions (Figure 3.9) mean a decreasing 'shading' effect and therefore enhanced warming. Since both CO_2 and SO_2 emissions are generally lowest in the sustainable-development scenarios, these two effects practically cancel each other out. This explains why the trajectories of temperature change during the first half of the 21st century are so similar in all groups of scenarios. Only after 2050 do CO_2 emissions begin to differ substantially between scenarios, and SO_2 emissions reach relatively low levels in most scenarios.

In contrast to the other scenarios, the coal-intensive A1C and A2 scenarios (and to a lesser extent also their mitigation counterparts) feature a substantial increase of global SO_2 emissions until 2040, but they then decline rapidly (Figure 3.9). Still sulphur emissions produce a strong cooling effect in A1C and A2 until 2090. As a consequence, global mean temperature change for A1C and A2 is lower than in the other cases until about 2050, although CO_2 emissions in the A1C and A2 baselines are the highest among all the scenarios examined here. However, on balance, this is not very good environmental news, since SO_2 emissions in A1C and A2 would have regionally catastrophic acidification impacts, in particular in Asia. Furthermore, SO_2 and NO_X emissions on levels suggested by the A1C scenario would pose serious threats to human health and world food security.

In view of these dismal environmental consequences, we consider A1C and A2 undesirable outcomes in practical terms. In our opinion, these two scenarios should be considered as 'if–then' exercises that prove that energy strategies heading that way may produce severe environmental pressures that will show the necessity at least to introduce policies that aim at implementing the mitigation variants of these scenarios. Better, of course, would be to avoid such developments from the beginning and implement policies that aim at the realization of one of the environmentally more compatible scenarios of our sets.

From the perspective of the 'precautionary principle', the considerable uncertainty around best-guess values means that even comparatively low levels of GHG emissions might have severe impacts on the climate and should therefore be avoided. The opposite perspective suggests that even carbon emissions near a dynamics-as-usual path may not lead to 'dangerous interference' with the climate and that costly mitigation efforts should therefore be avoided. One important conclusion from the big difference between these two conclusions is the need to avoid any possible confusion from the beginning and to aim for the more general goal of sustainable development rather than the narrower goal of climate stabilization.

3.5 ANALYSIS OF MITIGATION SCENARIOS

In this section, we present typical patterns of carbon abatement that arise if cumulative carbon emissions in MESSAGE are constrained in such a way as to stabilize the atmospheric CO_2 concentration during the 21st century. IIASA's scenarios include cases with concentration limits of 450, 550, 650, and 750 ppm CO_2. Since we are here concerned with typical patterns, we limit the discussion to presenting the scenarios with a concentration limit of 550 ppmv, the most common concentration used in mitigation analyses.

Figure 3.12 shows the CO_2 emissions trajectories of the four mitigation scenarios included here, comparing them with the respective baseline scenarios. The four trajectories are characterized by peaks of approximately 10 (B2) to 15 (A1C) GtC around the middle of the 21st century. After 2050, emissions decline to slightly below the 1990 level (6 GtC) by 2100 in all four scenarios. Note that the trajectories in Figure 3.12, following Riahi and Roehrl (2000b), are close to other emissions trajectories for 550 ppmv stabilization cases found in the literature. (See, for instance, Wigley *et al.*, 1996; Roehrl and Riahi, 2000). This means that 'flexibility in time', that is, the possibility, in MESSAGE, to meet the constraint of cumulative emissions without constraints for each point in time, does not make a major difference with respect to the timing of mitigation measures.

Emissions in the stabilization runs and their baseline counterparts do not differ significantly before the year 2020. After 2020, emission reductions become much more pronounced. This is partly because power plants have lifetimes on the order of 30–40 years, which makes for slow turnover in the energy capital stock, and partly because of the temporal flexibility built into the concentration constraint. The longevity of the capital stock means that the flexibility in time cannot be used to conclude that, prior to 2020, there would be no need for action to mitigate carbon emissions. Rather, we interpret the results shown in Figure 3.12 as the picture that arises if and when policy makers around the globe begin to aim for climate stabilization today.

Clearly, to meet a given CO_2 concentration stabilization level, more CO_2 has to be reduced in baseline scenarios with higher emissions than in those with lower emissions. However, costs and other efforts to reduce carbon emissions are determined not only by the baseline scenario's CO_2 emission levels but also by assumptions on technological progress (cost reductions over time), resource availability and socioeconomic development. This observation explains the different shapes of the patterned areas in Figure 3.12, which illustrates the main sources of CO_2 reductions in the global energy systems of the mitigation scenarios, compared to the respective baseline scenarios.

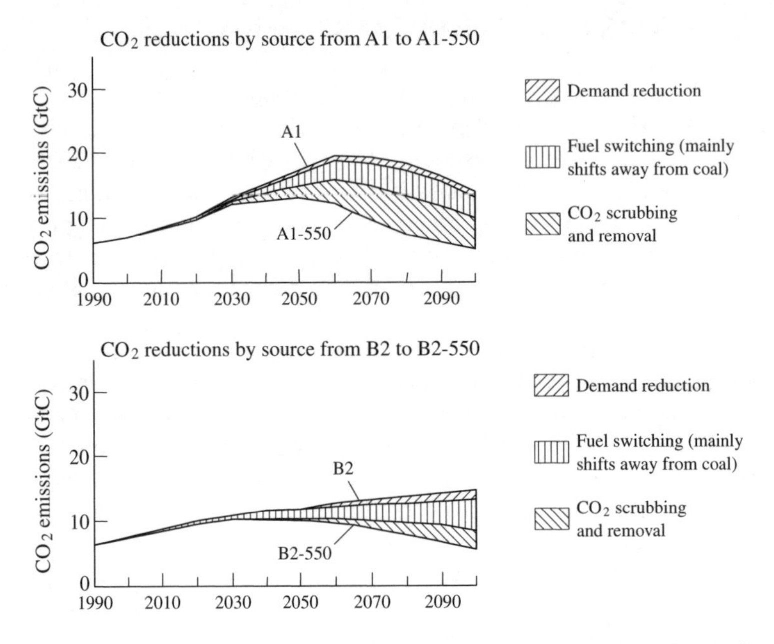

Note: The patterned areas depict the three main measures of CO_2 reductions in the mitigation scenarios.

Figure 3.12 *CO_2 emissions in four mitigation scenarios and their respective baselines*

The graphical presentation in Figure 3.12 disaggregates total mitigation relative to the respective reference case into the following three contributions:

- lower energy demand (enhanced energy conservation) due to higher energy costs of the stabilization case compared to the respective reference case.
- fuel switching away from carbon-intensive fuels such as coal;
- scrubbing and removing CO_2 in power plants and during the production of synthetic fuels, mainly methanol and hydrogen;

In the A2-550 and B2-550 scenarios, the largest contribution to emission reduction comes from structural changes in the energy system, primarily from replacing coal. To satisfy the carbon constraint, both of these two scenarios reduce coal's share of primary energy from 26 per cent in 1990 to 6 per cent (B2-550) and 17 per cent (A2-550) respectively by 2100. Of course, this is made possible by suitable assumptions on the overall availability of alternatives to coal.

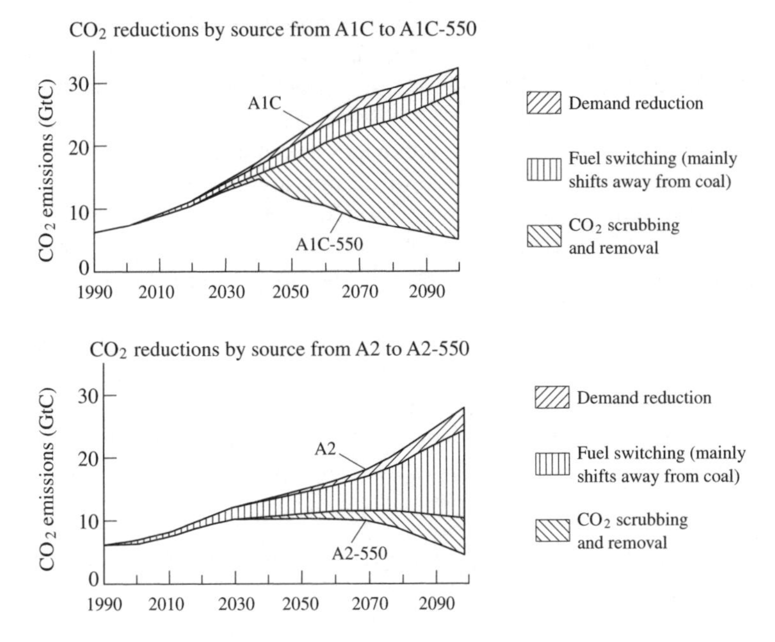

Figure 3.12 (continued)

In the A1C-550 scenario, CO_2 scrubbing is the major source of CO_2 emissions reduction. Almost 90 per cent of the emissions reductions in 2100 are due to scrubbing. Scrubbing occurs at the point of production of hydrogen and other synthetic fuels as well as in the electricity sector. By the same token, structural changes (fuel shifts) in the energy system (the principal contributor to the reduction in B2-550 and A2-550) play a minor role in A1C-550. This development is the consequence of high technology improvement rates in the coal sector (carbon capture and sequestration), which were already assumed for the A1C baseline. Even with the additional costs and electricity losses involved in CO_2 scrubbing, this option carries over as an attractive option in the A1C-550 mitigation scenario.

Carbon scrubbing is an important reduction source also in the A1-550 scenario, where more than 50 per cent of the CO_2 emissions reduction requirements are met by scrubbing. However, in contrast to A1C-550, where the main reason for the relatively high share of CO_2 scrubbing is the favourable cost development assumed for coal-based technologies, the main reason for the high share of scrubbing in A1-550 is the limited potential for structural changes from A1 to A1-550.

In the A1 baseline, technology cost reduction occurs relatively fast across all technologies, and coal is no exception. Consequently, in 2100, zero-carbon options already contribute more than 70 per cent to the energy supply in baseline scenario A1 (Figure 3.3). To meet the stabilization constraint, their share rises further, up to 78 per cent in A1-550, which is already the highest share among the stabilization scenarios (see Figure 3.3). Consequently, scrubbing and removal of carbon are needed in A1-550 to decarbonize the remaining fossil energy system, but this option accounts for no more than 12 per cent of the total primary share in this mitigation scenario. By 2100, 4.7 GtC per year are scrubbed and 2.1 GtC per year are reduced as a result of structural changes.

In all mitigation scenarios, price-induced energy demand reductions contribute to CO_2 emissions reduction. Of the four mitigation scenarios considered here, the A2-550 scenario shows the highest emissions reductions due to price-induced energy demand reductions (4 GtC in 2100), and A1-550 shows the smallest such effect (1 GtC in 2100).

3.6 ANALYSIS OF SUSTAINABLE-DEVELOPMENT SCENARIOS

We describe below the seven IIASA sustainable-development (SD) scenarios in relation to the following two scenario groups: all 34 IIASA scenarios, and the 400 SRES database scenarios. The idea behind doing this is to embed the ranges of the important descriptors of SD scenarios into ranges of larger sets of scenarios, thereby characterizing sustainable development.

3.6.1 Overview

Low pollutant emissions are a necessary requirement for environmental sustainability, but fulfilling this requirement satisfies only one criterion of our definition of sustainable-development scenarios (see Chapter 1). As we have seen above (see Figure 3.8), the GHG emission trajectories of mitigation scenarios overlap with those of the sustainable-development scenarios. Let us now expand this comparison by having a closer look at the evaluation of all four criteria in the three scenario sets. The evaluation is summarized in Table 3.6.

The following subsections describe results of these seven scenarios from the perspective of the sustainable-development criteria and the driving forces.

3.6.2 Sustained Economic Growth

Our working definition of sustainable-development scenarios proposes to measure economic growth in terms of GDP per capita. We therefore divide the discussion of this criterion into two parts, population and GDP.

Population

Indirectly, population is one of the fundamental driving forces of future emissions. Today there are three main research groups that project global population: the United Nations (UN, 1998), the World Bank (Bos and Vu, 1994) and IIASA (Lutz *et al.*, 1996, 1997). These three are the sources for the population projections of virtually all emissions scenarios included in the SRES database. Population is an exogenous input to the majority of models used to formulate the emissions scenarios of the SRES database. Only in exceptional cases is there a feedback from the E3 system on population dynamics included in the scenario.

Figure 3.13 shows the global population range of 46 SRES database scenarios (those that included projections of global population) and the 34 IIASA scenarios. The range for the SRES database scenarios is from more than 6 to about 19 billion people in 2100, with the central or median estimates in the range of about 11 billion.

The average long-term historical population growth rate has been approximately 1 per cent per year during the last two centuries and approximately 1.3 per cent per year since 1900. Currently, the world's population increases by about 2 per cent per year. The scenarios and other global population projections envision a slowing population growth in the future. The most recent doubling of the world population took approximately 40 years. Even the highest population projections in Figure 3.13 require 70 years or more for the next doubling while roughly half of the scenarios do not double population during the whole of the 21st century. The lowest average annual population growth (between 1990 and 2100) across all projections is 0.1 per cent, the highest is 1.2 per cent, and the median is approximately 0.7 per cent. As a result, global population in 2100 varies by factors of between one and less than four relative to 1990.

In all sustainable-development scenarios, population stabilizes during the 21st century at levels of around (or below) 11 billion people. This means that the global 'demographic fertility transition' is achieved in all SD scenarios. Demographic fertility transition was described by Easterlin (1978) as the effect of decreasing fertility rates of societies as a consequence of per capita incomes growing to a level around US$20 000 or higher. Despite these wide ranges among alternative global population projections, the

Table 3.6 Categorization of seven IIASA sustainable-development scenarios in comparison to typical high-impact and mitigation scenarios

	Criteria/ scenario	Sustainable economic growth	Equity ('significant improvement', balanced)		Low or no long-term environmental stress		Low or no medium-term environmental stress	
			Interregional	Intergenerational	GHGs	Scrubbing	Landuse	Acidification
Sustainable development	A1T	Yes	Yes	Yes	Yes	Yes	(Yes)	Yes
	B1	Yes	Yes	Yes	Yes	Yes	Yes	Yes
	B1T	Yes	Yes	Yes	Yes	Yes	Yes	Yes
	B1G	Yes	Yes	Yes	Yes	Yes	Yes	Yes
	WEC-A3	Yes	(Yes)	Yes	Yes	Yes	Yes	Yes
	WEC-C1	Yes	(Yes)	Yes	Yes	Yes	Yes	Yes
	WEC-C2	Yes	(Yes)	Yes	Yes	Yes	Yes	Yes
Non-sustainable baselines	A1B	Yes	Yes	No	Yes	Yes	No	Yes
	A1C	Yes	Yes	No	No	Yes	No	No
	A1G	Yes	Yes	No	No	Yes	No	No
	A2	Yes	No	No	No	Yes	(No)	No
	B2	Yes	(Yes)	No	No	Yes	(Yes)	Yes
	WEC-A1	Yes	(Yes)	No	No	Yes	(Yes)	(No)
	WEC-A2	Yes	(Yes)	No	No	Yes	(Yes)	No
	WEC-B	Yes	No	No	(No)	Yes	(Yes)	No

Stabilization/ mitigation scenarios	A1T-550	Yes	Yes	Yes	Yes	(Yes)	(Yes)	Yes
	A1B-550	Yes	Yes	No	Yes	No	No	Yes
	A1C-550	Yes	Yes	No	Yes	No	No	No
	A1G-550	Yes	Yes	No	Yes	No	No	No
	B2-550	Yes	(Yes)	No	Yes	(Yes)	(Yes)	Yes
	A2-550	Yes	No	No	Yes	No	(No)	No
	Analogous for the other stabilization scenarios*	Yes	Yes/No	Yes/No	Yes	Yes/No	Yes/No	Yes/No

Notes: A 'Yes' entry for acidification means that sulphur emissions in this scenario do not lead to sulphur depositions that exceed critical loads. '(Yes)' and '(No)' indicate borderline cases.
A 'Yes' entry for land-use change means that a combination of agricultural productivity increase and low population yields sufficient food supply with sustainable land management.
*14 scenarios stabilizing atmospheric carbon concentrations between 450 and 750ppmv. Note that, of the 20 mitigation scenarios, only two (A1T-550 and A1T-450) would qualify to be included in the group of sustainable-development scenarios.

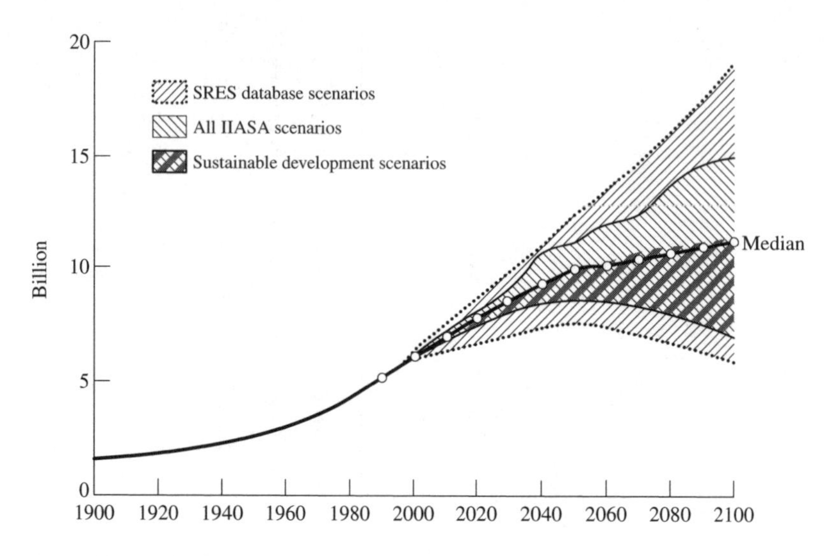

Note: 46 scenarios from the SRES database, 1990–2100.

Sources: Durand (1967), Demeny (1990), UN (1996); database: Morita and Lee (1998).

Figure 3.13 Global population, actual development, 1900–1990 and in the scenarios

range of this variable relative to the base year is the smallest of all driving forces considered in this comparison.

The 34 IIASA scenarios cover almost the whole range of population projections reported in the database. Most notably, however, the highest population growth assumed for the sustainable-development scenarios (11.7 billion in 2100) is roughly the same as the median from all the scenarios in the SRES database. Although the causal direction cannot be derived from the graph unambiguously, we conclude that slow population growth – if not stabilization of population at median levels – is a prerequisite for sustainable development.

Income and GDP growth

It appears unrealistic to assume that the developing regions can grow while the economies in the rest of the world stagnate. Closing the income gap between developing and industrialized world regions therefore means that per capita GDP in today's developing regions has to increase steeply. It thus follows that global economic development proceeds fast in sustainable-development scenarios. Figure 3.14 shows the future GDP per capita compared with actual development since 1950.

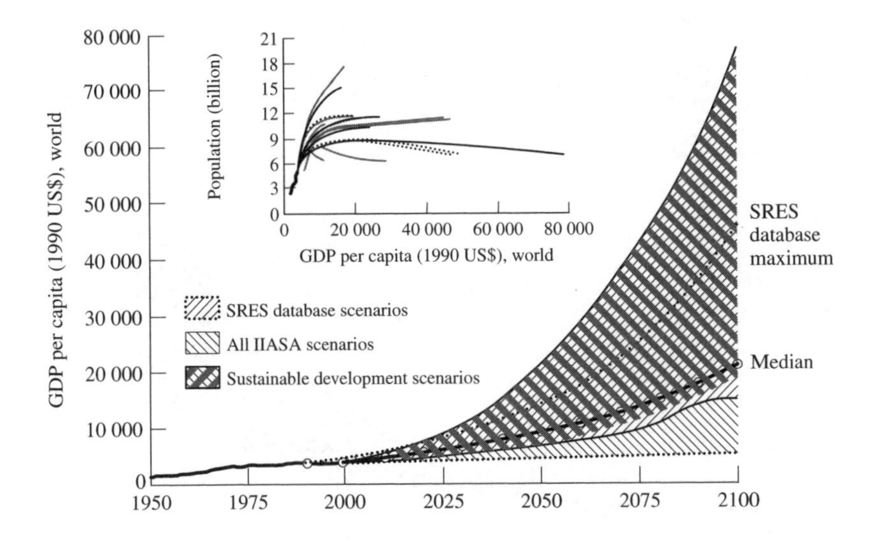

Notes: 28 scenarios from the SRES database, 1990–2100. The insert shows the development of global population as a function of GDP per capita. The gray lines in the insert depict the SRES database scenarios; the dotted lines show the IIASA sustainable-development scenarios, and the black lines show other IIASA scenarios. Historical data: UN (1993a, 1993b); database: Morita and Lee (1998).

Figure 3.14 Global average gross domestic product (GDP) per capita, 1950–90 and in scenarios for 1990–2100

Globally, GDP per capita has grown at an average annual growth rate (AAGR) of approximately 2 per cent since 1950. In the SRES database scenarios, the AAGRs between 1990 and 2100 of this variable range from 0.4 per cent per year to 2.2 per cent, with the median at 1.6 per cent. Relative to 1990, per capita GDP in 2100 increases by factors of between 1.5 and more than 12 for the SRES database scenarios. For comparison, average GDP per capita growth rates in the IIASA SD scenarios range from 1.4 to 2.7 per cent per year, which corresponds to a GDP per capita increase of 5 to 20 times between 1990 and 2100.

The insert in Figure 3.14 illustrates the relationship between population and GDP per capita in the scenarios. The insert shows that very high and very low population profiles lead to low income (GDP per capita) in 2100. Highest income is seen in scenarios where population growth is 'flat'.

From GDP per capita and population, total economic output is calculated by straightforward multiplication. Still, we present projections of aggregated global GDP separately (in Figure 3.15). Doing so also has a methodological aspect because the range of SRES scenarios used for

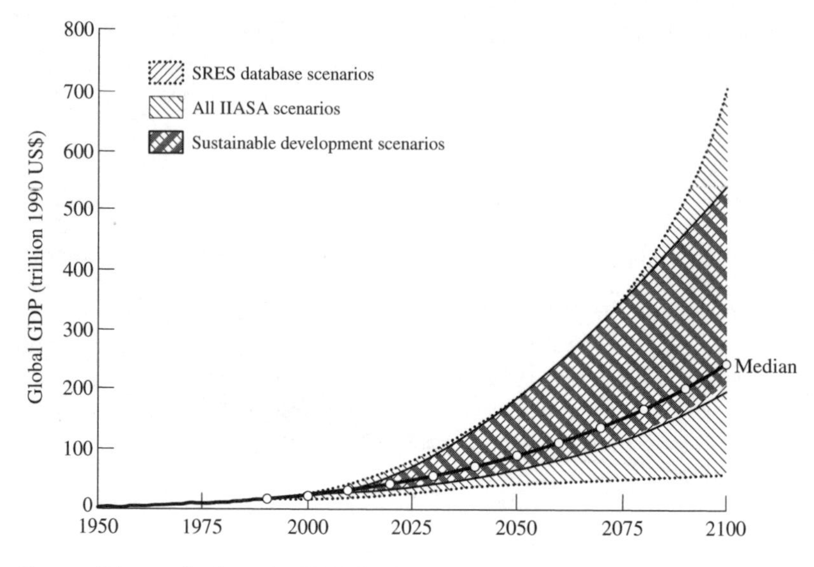

Notes: 193 scenarios from the SRES database, from 1990–2100. Historical data: UN (1993a, 1993b); database: Morita and Lee (1998).

Figure 3.15 Global economic product (GDP) in trillion (10^{12}) US$ (1990), development, 1950–90 and in the scenarios

generating the graph is based on no less than 193 database entries. This is a large sample when compared with the sample size of GDP per capita projections above. This difference stems from the fact that only a few scenarios in the database report both GDP and population projections.

Between 1950 and 1990, global GDP has grown by about 4 per cent per year. In the SRES database scenarios, the average growth rates between 1990 and 2100 range from 1.1 per cent per year to 3.2 per cent per year, with the median value at 2.3 per cent. The corresponding range of global GDP in the year 2100 is from about US$65 to more than US$700 trillion (with the median GDP of US$250 trillion) compared to a range of US$220 to US$550 trillion for the IIASA SD scenarios. Of the SRES database scenarios 90 per cent project global GDP values in 2100 that are between US$180 and 380 trillion.

The range of global GDP projections for the sustainable-development scenarios is practically identical to the one for the total IIASA scenario sample. This suggests that future *total* GDP levels are less critical in answering the question whether a scenario describes a sustainable future pathway or not. In other words, high total GDP alone is no guarantee for reducing economic and environmental inequities among world regions.

3.6.3 Intragenerational Equity: A Brief Methodological Digression

Sustainable-development scenarios illustrate developments that reduce inequity between world regions (intragenerational inequity). In our working definition of sustainable development, intragenerational equity is expressed as the ratios of GDP per capita between countries or world regions. At present, the ratio between the income in developing countries (ASIA + ROW) and developed countries (OECD + REFS) is approximately 6 per cent, a relatively low figure compared to the long-term ratios in the IIASA SD scenarios, which range from 21 to 64 per cent in 2100. These ratios are the result of very normative assumptions, in particular on a reversal of current trends through institutional, technical and financial transfers. In order to give readers a basis for applying their own judgment on how realistic these projections of future income growth are, we now document, in a small methodological digression, our own path from understanding past income growth to projecting its development in SD scenarios.

In short, we began by analysing the relation between the country's per capita GDP growth and its per capita income during the last four decades. On the basis of this analysis, patterns of the most successful country groups were identified and then compared to the future pathways of economic development in the sustainable-development scenarios. Figure 3.16 illustrates long-term average annual (per capita) economic growth rates for 88 countries for the period 1960–97, as a function of their per capita income.[16] The figure shows irregularly distributed GDP per capita growth rates across the whole income range indicating the very heterogeneous economic development of the countries in the last four decades. Clearly, no obvious pattern can be recognized from this picture.

Trying out several groupings of countries and several possibilities for the dependent variable (income) we find that using average annual growth rates of per capita GDP for 20-year periods, and looking specifically at Asian and OECD countries, drastically changes the apparent irregularity of the original picture. We show our results in Figure 3.17[17] (Asia) and Figure 3.18 (OECD), each time using time series for average (per capita) economic growth rates between 1960 and 1997.

Both figures show a clear pattern of a country's economic growth as a function of its per capita income. In particular, Figure 3.17 shows that the per capita GDP growth rate in most Asian countries follows (with a small number of exceptions) an 'inverse U'-shaped path; countries with an average annual per capita income in the neighbourhood of 200 1995 US$, such as Nepal or Bangladesh, show relatively low GDP growth rates.

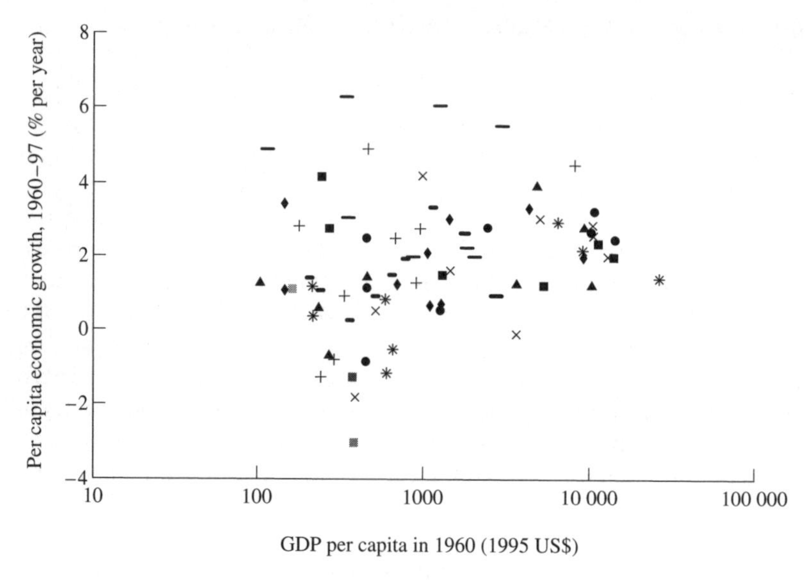

Source: World Bank (2000).

Figure 3.16 Long-term average annual economic growth in 88 countries (1960–97) and GDP per capita in 1960

With increasing GDP per capita, income growth rates tend to increase until annual per capita income reaches approximately 7000 (1995 US$). This was the case, for example, for the so-called Asian 'tiger economies' (Hong Kong, Korea, Singapore, Taiwan), which went through a phase of accelerated economic growth during the time period covered by the data points in this figure. Finally, for those countries in which GDP per capita increased beyond 7000 (1995 US$)/yr (for example, Japan, Singapore and Hong Kong) this trend changes, and the per capita GDP growth rates decline with further increase of per capita income.

This pattern of income growth is corroborated by Figure 3.18, which shows decreasing economic growth rates in OECD countries. There, the value at which economic growth rates begin to decrease is again approximately 7000 (1995 US$)/yr. According to Barro (1997), this pattern of economic growth reflects common phases of industrialization in different countries. The first phase, during early stages of industrialization, is characterized by accelerated economic development as a result of increasing demand for the build-up of new infrastructures. With increasing wealth, consumer demand increases too, and this in turn further accelerates economic growth. This trend continues until the shift from a pre-industrialized

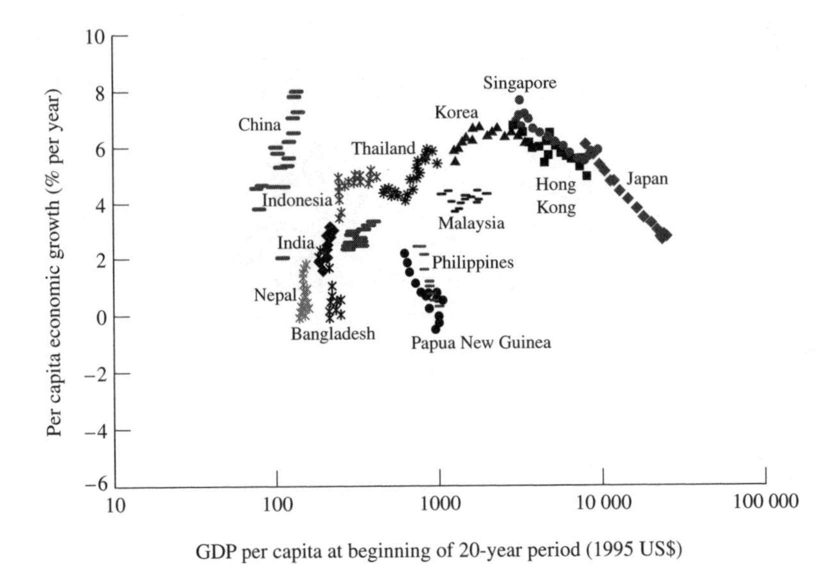

Source: World Bank (2000).

Figure 3.17 Long-term (20-year) average annual per capita economic growth rates in Asian economies 1960–97, and GDP per capita at the beginning of any of these 20-year periods

society to an industrialized economy is completed. The further a country develops and the closer it gets to the 'productivity frontier' (today represented by the OECD), the harder it is to increase productivity further. This results in decreasing economic growth.

For the later use of this pattern for long-term projections of economic growth, it appears particularly noteworthy that the 'inverse U' stretches over several orders of magnitude of income. This makes plausible the assumption that countries with lower income will follow the same path as other countries before. Of course, some optimism is required to project such an achievement for all low-income countries, but this plausibility argument led us to project income developments in the IIASA sustainable-development scenarios as shown in Figure 3.19.

The figure shows a much more distinct 'inverse U' in the more recent SRES scenarios than in the older IIASA-WEC scenarios. In the SRES scenarios, this leads to a readily visible narrowing of the 'income gap' between the developing regions and the OECD countries (note the difference between per capita incomes in 1990 – the left ends of the curves – and per capita incomes in the year 2100 – the right ends of the curves).

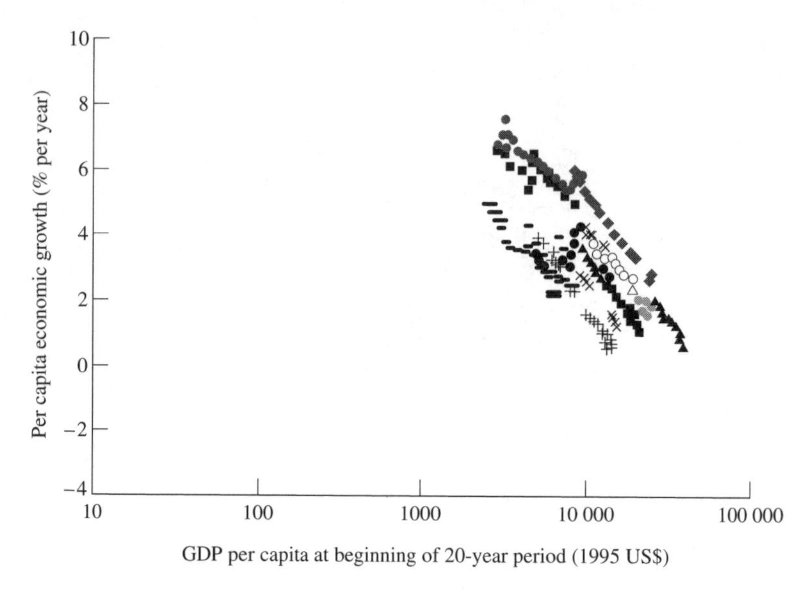

Source: World Bank (2000).

Figure 3.18 Long-term (20-year) average annual per capita economic growth rates in OECD countries, 1960–97, and GDP per capita at the beginning of any of these 20-year periods

Figure 3.19 does not show ranges of growth rates for the OECD region. We just note that these growth rates for the sustainable-development scenarios are the same as for the rest of SRES scenarios. This means that, under the assumptions made in the overall scenario set, sustainable development does not impede economic growth in the OECD region.

As we have commented in other places in this book, we want to emphasize that, in our opinion, the narrowing of the 'income gap' requires the world to embark on a sustainable-development path sooner rather than later. The later such a path is chosen, the higher the costs of reaching the same goal within the same time span and the more likely they can prove prohibitive. This suggests that there is little time to lose even in the near future if the long-term goals included in our scenarios are to be pursued.

3.6.4 Intergenerational Equity

Intergenerational equity, to leave future generations endowed with enough means to meet their own needs, is at the core of sustainability. In sustain-

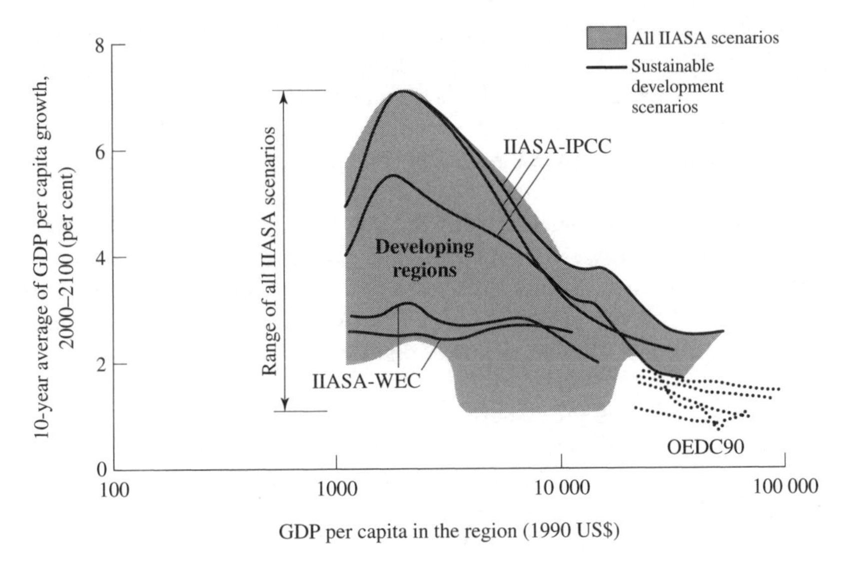

Note: OECD90 means OECD countries as of 1990.

Figure 3.19 Annual GDP per capita growth (10-year average) as a function of GDP per capita

able-development E3 scenarios, it is mainly the sustainable use of energy resources which is crucial for sustainability with respect to this criterion. Owing to its very nature, intergenerational equity can be studied reasonably only if the time frame of the study includes at least two generations. The IIASA E3 scenarios, covering the period between 1990 and 2100, therefore provide an appropriate frame for the analysis of intergenerational equity.

The parameters and variables most relevant for the assessment of intergenerational equity in long-term E3 scenarios such as IIASA's are the discount rate, the use of hydrocarbon resources and energy intensity. Let us discuss each of them in more detail now.

Discount rate

Discussing intergenerational equity in a quantitative way inevitably involves comparing costs and benefits that materialize at different points in time. Doing so usually involves the use of a discount rate, which is also a measure of time preference. Consequently, discount rates used in long-term environmental studies that include impacts on future generations are usually lower than those found in empirical studies of individual consumer behaviour. The latter include statistical studies of individual decisions, such as savings decisions, choices in financial markets (Thaler, 1981; Lowenstein and Thaler, 1989) and the valuation of public projects (Cropper *et al.*, 1994) as well as psychological research (Lowenstein and Elster, 1992).

Recently, variable (so-called 'hyperbolic') discounting has been proposed for sustainable-development research. The concept was even supported also by practical experimental evidence, but this evidence did not remain unquestioned. For a brief survey of the literature on hyperbolic discounting and some critique, see Fernandez-Villaverde and Mukherji (2002).

The IIASA model assumes a discount rate of 5 per cent. This reflects the fact that the typical projections in the IIASA model extend 30 to 50 years (that is, the average lifetime of power plants).

Hydrocarbon reserves and resources

One key issue in the debate on sustainability is the question whether current extraction rates of mineral resources jeopardize the 'ability of future generations to meet their own needs'. In particular, it is often argued that the fossil fuels oil, gas and coal are being extracted in a non-sustainable way. During the oil crises of the 1970s and 1980s, it was frequently feared that the world could run out of fossil energy in the near future. The long-term historical evidence (for example, Rogner, 1997; Nakićenović *et al.*, 1998), however, suggests that the situation is less drastic than then feared, mainly owing to insufficient appreciation of the difference between reserves and resources. Reserves are defined as only that part of the resources that it is technically and economically feasible to extract (see section 2.2 above). The concept of reserves is therefore a dynamic one, depending on the available technologies and the market conditions at each point in time.

In the IIASA model runs, the physical resource base (that is, reserves plus resources) is assumed to be identical in all scenarios. However, the assumed quantities of available reserves depend strongly on the assumptions of the then technoeconomic situation assumed to prevail in a scenario. This is the key to creating the possibility that the reserve-to-production (R/P) ratio of fossil fuels is the same in 2100 as in 1990. To make this view more plausible, let us look into the development of the R/P ratios of oil and natural gas during the past decades (Figure 3.20).

The figure shows that technological progress and changing market conditions have led to comparable R/P ratios for crude oil and natural gas since the 1970s at the global level (not at the country level of course). During the 1980s, the reserve-to-production ratio for oil and gas even increased as resources were continuously transformed into reserves (BP Amoco, 1999).

With this piece of hindsight, let us now look at the long-term R/P ratios (for gas and oil for the year 2100) for the sustainable-development scenarios of the SRES family (Table 3.7).[18] Scenario estimates for the year 2100 are

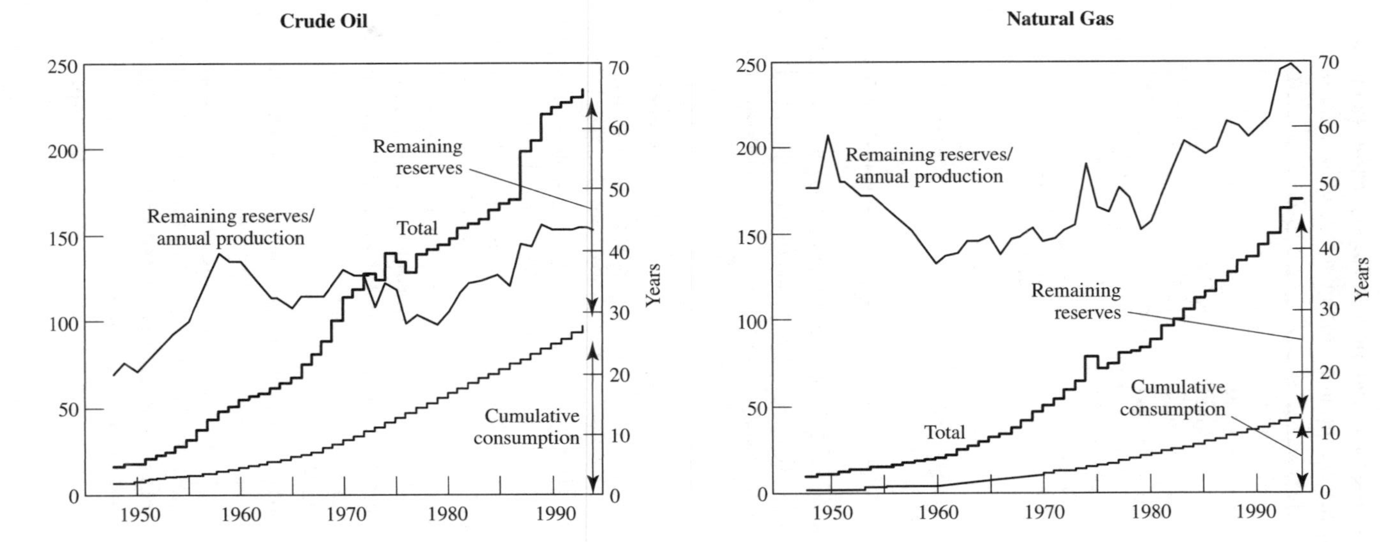

Source: Nakićenović *et al.* (1998).

Figure 3.20 Development of global oil and gas reserves, billion (10^9) tons of oil equivalent, 1950–95

Table 3.7 Reserve-to-production ratio and global resource consumption of natural gas and oil in sustainable-development scenarios

	Natural gas		Oil	
	Global consumption (EJ/yr)	R/P ratio (years)	Global consumption (EJ/yr)	R/P ratio (years)
SD scenario estimates for 2100				
SRES-A1T	196	127	77	178
SRES-B1	215	49	45	55
SRES-B1G	244	40	53	44
SRES-B1T	166	81	48	54
1990 value	*72*	*58*	*139*	*43*

compared to values for 1990. The long-term R/P ratio for the SD scenarios either increases or stays roughly at 1990 levels. Compared to 1990, gas consumption gains importance and increases, while oil consumption decreases significantly across all SD scenarios.

The high values of consumption of natural gas, combined with high R/P ratios (Table 3.7) show the relative abundance (compared to oil) of affordable natural gas in the SD scenarios in the 21st century. This abundance is based mainly on the assumption of rapid technological progress that will allow the extraction of vast amounts of non-conventional gas, including methane hydrates (Rogner, 1997) in a cost-effective way.[19] Further, the abundance of cheap coal appears to constitute no threat to intergenerational equity (ibid.). No exhaustion of world coal reserves is in sight within the next few hundred years. All in all, owing to the abundance of cheap coal and the possibility of natural gas being a virtually renewable energy source, we argue that environmental sustainability is much more an issue for sustainable development than limited fossil-fuel resources threatening intergenerational equity.

In this regard, possible negative effects on the environment include climate change, land use change and the possible environmental impacts of extracting methane hydrates. Whereas the effects of fossil fuel use on climate change have been analysed in much detail, still very little is known about the sustainability of the large-scale extraction of methane hydrates (in terms of potential costs and environmental impacts).

Primary-energy intensity

Rather than focusing solely on mitigating the negative consequences of energy use, for example, by shifting from fossil energy to non-fossils to mitigate GHG emissions, strategies aiming at sustainable development will have to go further and also attempt to reduce energy use to begin with, for example, by limiting the material flows in an economy. Expressing such a goal in aggregate terms, this means strategies aiming to reduce the primary-energy intensity; that is, the amounts of primary energy used per unit of GDP. Of the scenario driving forces considered here, we consider primary-energy intensity of GDP as the most descriptive of technological progress in the energy system. In all scenarios, global economic growth outpaces the increase in global energy consumption, leading to substantial reductions in the ratio of primary-energy consumption per unit of GDP. Typically, higher GDP growth rates at any point in time and in any world region correspond to faster decline rates of energy intensity because, during periods of fast economic growth, inefficient technologies are jettisoned faster in favour of more efficient ones. Also the structure of the energy system and patterns of energy services change faster during high-growth periods, having the same effect on the primary-energy intensity of GDP. This mechanism is of course most important in the sustainable-development scenarios, where energy intensities are reduced significantly compared to the other scenarios (Figure 3.21).

The figure shows the relationship between energy intensity and GDP per capita in the scenarios and the actual development between 1960 and 1990. As can be seen, increasing wealth is in general projected to be associated with lower energy intensity at the global level. The lowest energy intensity improvement rate of the database scenarios is 0.6 per cent per year and the highest is 1.8 per cent per year, with the median at about 1 per cent per year, corresponding to the long-term historical trend. The subset of SD scenarios shows a very clear picture: practically the entire range of projections falls below the median rates of the SRES database.

3.6.5 Long-term Environmental Stress: GHG Emissions and Climate Change

Carbon intensity

Although average annual decarbonization of the world's primary-energy supply has been not more than 0.3 per cent (Figure 3.22), the trend has persisted throughout the last two centuries (Nakićenović, 1996). The overall tendency towards lower carbon intensities is due to the continuous replacement of fuels with higher carbon content by those with lower carbon

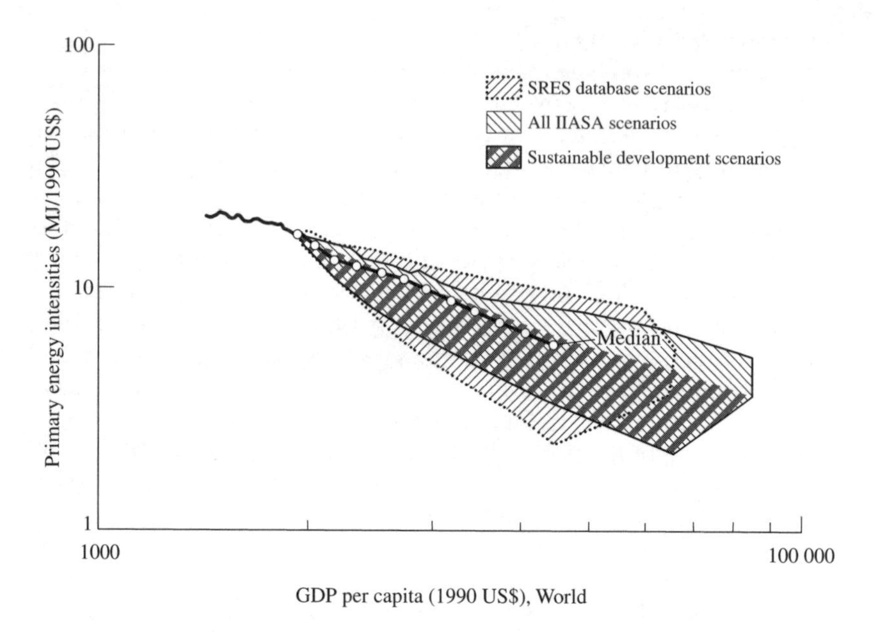

Notes: 28 scenarios from SRES database, 1990–2100; historical data: IEA (1993), *World Development Report* (1993); database: Morita and Lee (1998).

Figure 3.21 Global primary-energy intensity of GDP in relation to GDP per capita on logarithmic scales, development 1960–90 and in the scenarios

content, such as the replacement of coal by natural gas or non-fossil energy sources. Note that net carbon emissions are used (that is, bioenergy is excluded) to calculate the carbon intensities. The differences in the carbon intensities for the base year are explained by different accounting conventions and data problems in general.

Figure 3.22 shows the projections of carbon intensity of the SRES scenarios and the actual development since 1900. The highest projected average decarbonization rates in the database are near 3.3 per cent per year between 1990 and 2100. This means a reduction of energy-related carbon emissions per unit of energy by a factor of 40 over this time horizon, which, in turn, leads to an energy system with almost no carbon emissions. In the IIASA scenarios, decarbonization is fastest in the SD scenarios and those mitigation scenarios that stabilize atmospheric carbon concentrations at 550ppmv or below. Their respective ranges of projected carbon intensities for the year 2100 are from 2.4 kgC/MJ to 6.3 gC/MJ for SD scenarios and 0.8 gC/MJ to 4.5 gC/MJ for the mitigation scenarios.

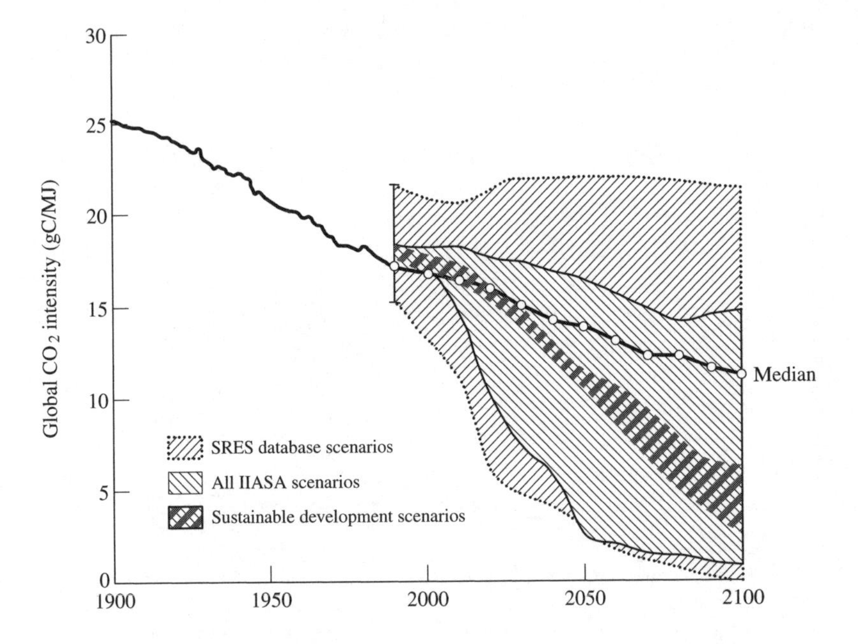

Notes: 140 scenarios from SRES database, 1990–2100; historical data: Nakićenović (1996); database: Morita and Lee (1998).

Figure 3.22 Global carbon intensity of primary energy, actual development, 1900–1990 and in the scenarios

Much more so than for energy intensity, carbon intensity in the SD scenarios is thus much lower than in the other scenarios, decarbonization rates in most cases falling to less than half the median value.

CO_2 emissions

The range of projected carbon dioxide emissions across all scenarios in the SRES database is indeed large, ranging, in 2100, from ten times the current emissions all the way down to negative net emissions: that is, a situation of carbon sinks more than outweighing energy-related sources, assumed in some scenarios. There are many possible explanations for this wide range, including many that are plausible. The most important explanation is the high uncertainty surrounding the evolvement of the main driving forces during the 21st century. Another factor to consider is that, implicit in some scenarios with particularly high carbon emissions, is the assumption that climate change will prove less of a problem than is feared by most scientists today.

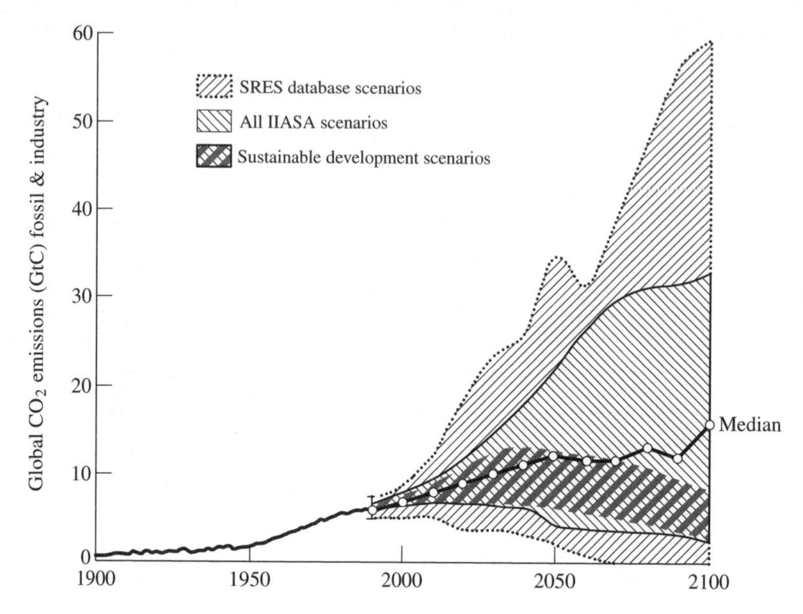

Notes: 140 scenarios from SRES database, 1990–2100; historical data: Marland *et al.* (1994); database: Morita and Lee (1998).

Figure 3.23 Global carbon dioxide emissions, actual development, 1900–1990 and in the scenarios

Figure 3.23 shows the global CO_2 emission paths from 1990 to 2100 for the scenarios and the actual emissions from 1900 to 1990. This figure again shows the wide differences that can be found in the base year data from the database owing to methodical differences among the scenarios (for example, different data sources, definitions and so on) and to different base years assumed in the analysis.[20] According to Marland *et al.* (1994), global CO_2 emissions from energy production and use (thus excluding industrial emissions such as those from cement production) in 1990 were estimated at 5.9 billion (10^9) tons of carbon (GtC). In comparison, the 1990 values in the scenarios reviewed range from 4.8 to 7.4 GtC with a median of 6.4 GtC. Consistent with Marland *et al.* (1994), the IIASA scenarios use a value of 6.2 GtC of carbon emissions from fossil fuel consumption and industrial sources (mainly cement production) in 1990, and land use-related emissions are excluded.

Global CO_2 emissions have increased at an average annual rate of about 1.7 per cent between 1900 and 1990. If this trend continued, global emissions would double by the year 2030, and many scenarios in the database describe exactly such a development. However, by as early as 2030, the range is very wide around this value of double global emissions. The highest projections by SRES database scenarios have emissions four times the 1990 level by 2030, while the lowest are barely above half the current emissions. This range keeps widening after 2030, the highest projected emissions including a tenfold increase over 1990 emissions by 2100 and, as mentioned, a lower end of negative net emissions. The median projection corresponds to global emissions of about 15.4 GtC in 2100 (about a threefold increase over 1990) and to more than a doubling of atmospheric CO_2 concentrations (approximately 750 ppmv) by 2100. A number of scenarios in the low range are consistent with stabilizing concentrations at levels of 450 ppmv, which are expected to prove relatively benign to the global climate.

The carbon emissions range for the IIASA scenarios in 2100 (2.3 GtC to 32.7 GtC) covers the emission trajectories of more than 95 per cent of all scenarios from the SRES database. Only a few 'outlier' projections included in the database fall outside the range given by the IIASA scenarios. In the IIASA scenarios, projected carbon emissions are of course lowest in the SD and mitigation scenarios that stabilize atmospheric CO_2 concentrations at 450 ppmv by 2100.

As shown in Figure 3.23, carbon emissions projected by SD scenarios range from 2.9 GtC to 8 GtC in 2100, leading to atmospheric CO_2 concentrations of between 550 and 650 ppmv in 2100. Again this range of emissions is well below the SRES median, but, in this case, it is important to note that this is not true for all time periods of the 21st century. Prior to 2050, we see the range of carbon emissions of SD scenarios reaching clearly above the median. This emphasizes the importance of the 'flexbility in time'; that is, carbon mitigation later in the century can still lead to sustainable development. However, this observation must not be mistaken for justifying inaction now. As we shall argue in more detail below, policy action towards sustainable development must begin as early as possible.

Climate implications

The SRES database does not include direct information on the scenarios' climate impacts such as global average temperature change. We therefore used the MAGICC model, version 2.3 (Wigley and Raper, 1997) to calculate global mean temperature change by the year 2100 relative to 1990 from annual emissions. This version of the model supports regionalized (three world regions) SO_2 emissions input data, which are needed to calculate the regionally different cooling effects of sulphate aerosols. For radiative forcing, the parameterizations reported in Myhre *et al.* (1998) were used. The other model input parameters for MAGICC used here are similar to those used by the IPCC in the Second Assessment Report (Houghton *et al.*, 1996).

We applied MAGICC using just the emission trajectories of the 34 IIASA scenarios, but since the range covered by these emission trajectories is representative of all SRES scenarios, the resulting ranges of temperature change are also representative.

Scenario-independent inputs into MAGICC include time series of anthropogenic sources of CO_2, CH_4, N_2O, SO_2, CFC/HFC/HCFC, PFC, SF_6, CO, VOCs and NO_X. Energy-related emissions of CO_2, CH_4 and SO_2 are direct outputs of MESSAGE. Non-energy sector emissions of CO_2, CH_4, N_2O, SO_2, CO, VOCs and NO_X were estimated with a spreadsheet model, using corresponding land-use change model runs with equivalent input assumptions from the AIM model (Jiang *et al.*, 2000; Riahi and Roehrl, 2000a). Data for PFCs, SF_6 and HFCs were taken from Fenhann (2000). The same publication was the source of emission data for ozone-depleting substances covered by the Montreal Protocol (for example, CFCs, HCFCs). Fenhann based his estimates on the Montreal Protocol scenario (A3) from the 1998 Scientific Assessment of Ozone Depletion (WMO/UNEP, 1998).

With these inputs, MAGICC calculates atmospheric GHG concentrations, global radiative forcing and temperature change. The scenarios' results in terms of global mean temperature change from 1990 to 2100 are illustrated in Figure 3.24, which shows that it takes several decades for expected global mean temperature changes in SD scenarios to develop in a way that is distinctly different from the 'dynamics-as-usual' SRES-B2 scenario. This slow response to SD policies is due to the inertia of both the energy system and the climate system. In 2100, best-guess temperature change in all IIASA scenarios varies from 1.3 to 3°C according to scenario. Again the stabilization cases and the SD scenarios cluster near or below the SRES-B2 scenario trajectory.

Note that the climate model uncertainties are very large indeed, even for a given GHG concentration and given radiative forcing. The best-guess values of temperature change across all IIASA scenarios span roughly the same range as the one introduced by a variation of the climate sensitivity parameter for the medium B2 scenario alone (shown as vertical bars in Figure 3.24). In other words, the compounded uncertainty surrounding the future development of socioeconomic, demographic and technological change assumed for the scenarios is roughly the same size as the uncertainty of the actual climate sensitivity. This emphasizes the importance of future research, which should focus on reducing the uncertainty of emissions scenarios as well as the uncertainties of the climate models.

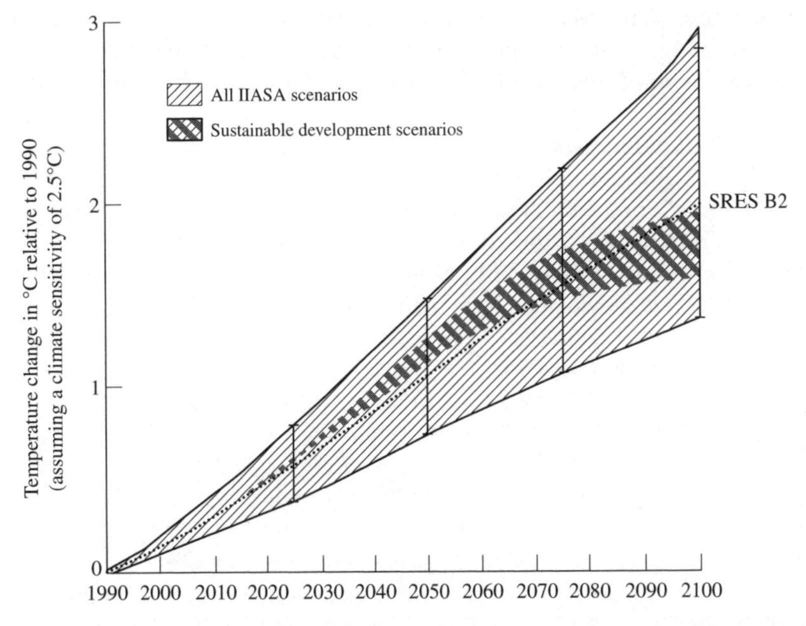

Notes: The grey area depicts the range for the 34 IIASA scenarios and the crosshatched area the range for SD scenarios assuming a median climate sensitivity of 2.5°C. The vertical bars indicate the uncertainty range for the 'dynamics-as-usual' SRES-B2 scenario arising from a variation of the climate sensitivity between 1.5 and 4°C (Watson *et al.*, 1996).

Figure 3.24 Average global mean temperature change relative to 1990, scenario projections 1990–2100

3.6.6 Acidification, Land Use

As we have illustrated above (Table 3.6), a scenario may fulfil the criterion of environmental sustainability at the global scale and from a very long-term perspective, but may at the same time be unsustainable at the local scale and in the more medium term. This possibility is illustrated by the SRES-A1B scenario, in which a rapid and wasteful development at a world regional level during the medium term prevents it from meeting the environmental criterion of our definition of SD scenarios.

To assess medium-term environmental stress in our scenarios, we examined two major kinds of environmental damage, land use change-related impacts and critical acidification loads, mainly caused by SO_X and NO_X emissions. For example, rapid energy demand increase in Asia in some A1 scenario variants leads to considerable sulphur emissions and critical acidification loads in that world region, even though new and clean technologies (for example, sulphur scrubbers) are assumed to be rapidly diffusing.

The medium-term effects of such a scenario may significantly threaten regional food security and public health. This possibility has been found very real in an earlier study involving IIASA's ECS Program, which analysed variants of the IIASA-WEC scenarios in terms of their critical loads in the Asian region (Nakićenović *et al.*, 1997). That study shows that, even though global sulphur emissions for all IIASA scenarios are rather close to the low range of the scenario literature (Figure 3.25), acidification impacts may have disastrous socioeconomic impacts at the local level, even though they do not significantly reduce global economic growth. For more detail, see Nakićenović *et al.* (1997) and McDonald (1999).

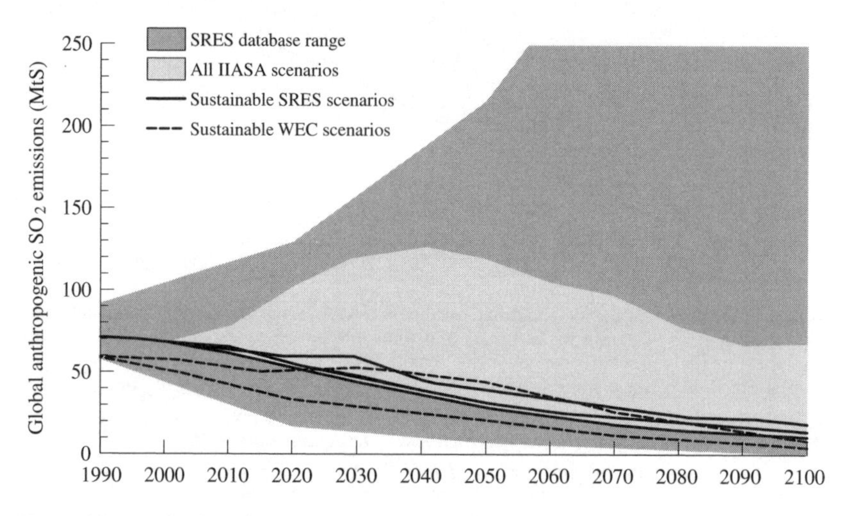

Note: 88 scenarios from SRES database, 1990–2100.

Figure 3.25 Global anthropogenic SO_2 emissions in the scenarios

The light grey area in the figure depicts the range for the 34 IIASA scenarios. SD scenarios from the SRES family are shown as uninterrupted lines, and SD scenarios from the IIASA-WEC study as dashed lines. Note that the IIASA-WEC scenarios do not include non-energy-related anthropogenic sulphur emissions. The database used is Morita and Lee (1998).

The most important land use impact of energy conversion and use goes back to (traditional as well as commercial) biomass use. See Fischer and Schrattenholzer (2001), who estimate global and world regional bioenergy potentials in consistency with land use changes. The IIASA SRES scenarios have used assumptions on bioenergy potentials that were consistent with those estimates.

NOTES

1. The EMF homepage (http://www.stanford.edu/group/EMF/home/index.htm) leads to a wealth of information on EMF studies.
2. The ranges of SD scenarios are so small that it does not seem worthwhile to include medians of SD scenarios as well.
3. Climate change scenarios were not considered during the preparation of the SRES because the task was to survey carbon emissions in scenarios that did not include explicit climate policies. Scenarios that included more general environmentally benign strategies such as sustainable development were admitted, however. This was not just a haphazard distinction because mitigation scenarios were dealt with later (in the Third Assessment Report of the IPCC, 2001). See also our description in the following subsection.
4. The rapid economic development in the A1 scenarios implies a replication of the post-World War II growth experience of Japan and South Korea or the recent economic development of China on a global scale. The global economy is projected to expand rapidly at an average annual rate of 3 per cent to 2100. Although this rate is roughly in line with long-term historical experience over the last 100 years, it should be noted that 100-year statistics are available essentially only for industrialized countries (Maddison, 1993).
5. For instance, non-fossil electricity (photovoltaics, advanced nuclear) may become available at costs of less than 10 US$/MWh).
6. Advanced nuclear power plants are defined as technologies that produce energy with higher efficiency and increased safety compared to today's nuclear standards. Their technological design is not pre-specified in the model. Advanced nuclear technologies should be interpreted as a generic rather than as a specific technology. Examples of specific technologies that would fit this description might include high-temperature reactors (producing hydrogen), fast-breeder reactors with new, modified designs, and other nuclear-fission technologies.
7. Also the gas-intensive A1G scenario has high future emissions (31 GtC in 2100). However, we do not describe a CO_2 mitigation case for the A1G baseline in this book. For more information on carbon mitigation in A1G, see Roehrl and Riahi (2000).
8. In addition to these 17 stabilization scenarios, IIASA-ECS has also developed two mitigation cases based on the sustainable-development scenario A1T.
9. Often the argument is made that it is implausible to assume the same kind of technological progress and stringent emission mitigation at the same time. That argument gave rise to the sustainable-development scenarios, which we describe separately in other parts of this book. Meanwhile we would argue that using the same availability of technologies also gives important insights, in particular because we use a price-responsive set of demands.
10. In particular, our model implies mechanisms such as unrestricted emissions 'trading' on a global scale, however without the specific modelling of these transactions or specific country or regional quotas, but rather a global quota.
11. Note that 550 ppmv level is special in that it was identified by many researchers as a realistic, achievable global target with possibly less-than-drastic climate impacts. Lower levels appear likely to be rather costly or assuming challenging policy changes ('sustainable development'). Levels much higher than 550 ppmv imply much higher probabilities of high impact climate change. As a result, the 550 ppmv level is often used by decision makers (e.g., the European Commission) for policy discussions. This has been essentially a political choice.
12. This note is the same as note 6. Advanced nuclear power plants are defined as technologies that produce energy with higher efficiency and increased safety compared to today's nuclear standards. Their technological design is not pre-specified in the model. Advanced nuclear technologies should be interpreted as a generic rather than as a specific technology. Examples of specific technologies that would fit this description might include high-temperature reactors (producing hydrogen), fast-breeder reactors with new, modified designs, and other nuclear-fission technologies.
13. We define technology clusters in a more formal way in the next chapter. For the purpose

of the description here, the given explanation suffices. Also, if we think there is no risk of confusion, we shall use the short term 'technology' instead of 'technology cluster' for simplicity.
14. For the description here, we define the interval between minimum and maximum market share of a technology as its 'range'.
15. The absolute amounts of CH_4 reductions in the stabilization scenarios depend strongly on the emissions level in the baseline (the lower bound of the range reflects reductions in the A1 scenario, the upper bound for the A2 scenario).
16. Note that the choice of the 88 countries used in the following analysis was purely driven by data availability. In some respects, one may argue that this is already a pre-selection of data in favour of the economically 'successful' countries.
17. This figure is identical with Figure 2.1. It is repeated here for easier reference.
18. The SD scenarios of IIASA-WEC had not paid specific attention to intergenerational equity. Assumptions on fossil-fuel reserves were formulated from a technical perspective only, that is, they just served the purpose of making the scenarios feasible in the spirit of the scenario storylines. As we have said above (pp. 54), the three IIASA-WEC scenarios A3, C1 and C2 are in a sense earlier versions of the SRES A1T, B1 and B1T scenarios. We therefore classified them as SD scenarios despite the impossibility of documenting suitable R/P ratios for these scenarios for the year 2100.
19. According to the 'Modern Russian–Ukrainian theory of abyssal, abiotic petroleum origins', oil and natural gas are produced abiogenically in the earth's core. Oil and natural gas could thus be virtually renewable energy sources. See Gold and Soter (1980), Krayushkin *et al.* (1994), and Odell (2000).
20. Also the possibility of errors cannot be excluded.

REFERENCES

Barro, R.J. (1997), *Determinants of Economic Growth*, Cambridge, MA: MIT Press.

Bos, E. and M.T. Vu (1994), *World Population Projections: Estimates and Projections with Related Demographic Statistics*, 1994–5 edn, Washington, DC: World Bank.

BP (British Petroleum) Amoco (1999), *BP Statistical Review of World Energy 1999*, London: BP.

Cropper, M.L., S.K. Aydede and P.R. Portney (1994), 'Preferences for life-saving programs: How the public discounts time and age', *Journal of Risk and Uncertainty*, **8**, 243–65.

Demeny, P. (1990), 'Population', in B.L.Turner *et al.* (ed.), *The Earth as Transformed by Human Action*, Cambridge: Cambridge University Press.

Durand, J.D. (1967), 'The modern expansion of world population', *Proceedings of the American Philosophical Society*, **111** (3), 136–59.

Easterlin, R.A. (1978), 'The economics and sociology of fertility: a synthesis', in C. Tilly (ed.), *Historical Studies of Changing Fertility*, Princeton: Princeton University Press, pp.571–613.

Fenhann, J. (2000), 'Industrial non-energy, non-CO_2 greenhouse gas emissions', *Technological Forecasting and Social Change*, **63** (2–3), 313–34.

Fernandez-Villaverde, J. and A. Mukherji (2002), 'Can we really observe hyperbolic discounting?', PIER Working Paper 02-008, Penn Institute for Economic Research, University of Pennsylvania, Philadelphia.

Fischer, G. and L. Schrattenholzer (2001), 'Global bioenergy potentials through 2050', *Biomass and Bioenergy*, **20** (3), 151–9.

Gold, T. and S. Soter (1980), 'The deep-earth-gas hypothesis', *Scientific American*, **242** (6), 154–61.

Grübler, A. (1998), 'Emissions scenarios database and review of scenarios', *Mitigation and Adaptation Strategies for Global Change*, **3** (2–4), 383–418.
Houghton, J.T., L.G. Meira Filho, B.A. Callander, N. Harris, A. Kattenberg, and K. Maskell (eds) (1996), *Climate Change 1995. The Science of Climate Change*, contribution of Working Group I to the Second Assessment Report of the Intergovernmental Panel on Climate Change, Cambridge: Cambridge University Press.
Houghton, J.T., L.G. Meira Filho, J. Bruce, Hoesung Lee, B.A. Callander, E. Haites, N. Harris and K. Maskell (eds) (1995), *Climate Change 1994: Radiative Forcing of Climate Change and an Evaluation of the IPCC IS92 Emissions Scenarios*, Cambridge: Cambridge University Press.
IEA (International Energy Agency) (1993), *Energy Statistics and Balances for OECD and Non-OECD countries, 1971–1991*, Paris: OECD.
Jiang, K., T. Masui, T. Morita and Y. Matsuoka (2000), 'Long-Term GHG emission scenarios for Asia–Pacific and the World', *Technological Forecasting and Social Change*, **63** (2–3), 207–30.
Kram, T., T. Morita, K. Riahi, R.A. Roehrl, S. van Rooijen, A. Sankovski and B. de Vries (2000), 'Global and regional greenhouse gas emissions scenarios', *Technological Forecasting and Social Change*, **63**, 335–71.
Krayushkin, V.A., T.I. Tchebanenko, V.P. Klochko, Ye. S. Dvoryanin and J.F. Kenney (1994), 'Recent applications of the modern theory of abiogenic hydrocarbon origins: drilling and development of oil & gas fields in the Dneiper-Donets Basin', VIIIth International Symposium on the Observation of the Continental Crust through Drilling, Sante Fe, NM, DOSECC: 21–4.
Lowenstein, G. and J. Elster (eds) (1992), *Choice over Time*, New York: Russell Sage Foundation.
Lowenstein, G. and R.H. Thaler (1989), 'Anomalies: intertemporal choice', *Journal of Economic Perspectives*, **3**, 181–93.
Lutz, W., W. Sanderson and S. Scherbov (1996), 'World population scenarios in the 21st century', in *The Future Population of the World: What Can We Assume Today?*, London: Earthscan, pp.361–96.
Lutz, W., W. Sanderson and S. Scherbov (1997), 'Doubling of world population unlikely', *Nature*, **387** (6635), 803–5.
Maddison, A. (1993), 'Monitoring the world economy 1820–1992', Development Centre of the Organisation for Economic Co-operation and Development, Paris.
Marchetti, C. and N. Nakićenović (1979), 'The dynamics of energy systems and the logistic substitution model', RR-79-13, IIASA, Laxenburg.
Marchetti, C. (1980), 'Society as a learning system: discovery, invention, and innovation cycles revisited', *Technological Forecasting and Social Change*, **18**, 267–82.
Marland, G., R.J. Andres and T.A. Boden (1994), 'Global, regional, and national CO_2 emissions', in T.A. Boden, D.P. Kaiser, R.J. Sepanski and F.W. Stoss (eds), *Trends '93: A Compendium of Data on Global Change*, ORNL/CDIAC-65, Carbon Dioxide Information Analysis Center, Oak Ridge National Laboratory, Oak Ridge, pp.505–84.
Marland, G., T.A. Boden, R.J. Andres, A.L. Brenkert and C. Johnston (1999), 'Global, regional and national CO_2 emissions', in *Trends: A Compendium of Data on Global Change*, Carbon Dioxide Information Analysis Center, Oak Ridge National Laboratory, US Department of Energy, Oak Ridge.
McDonald, A. (1999) 'Combating acid deposition and climate change', *Environment*, **41** (3).

Metz, B., O. Davidson, R. Swart and J. Pan (eds) (2001), *Climate Change 2001: Mitigation*, contribution of Working Group III to the Third Assessment Report of the Intergovernmental Panel on Climate Change, Cambridge: Cambridge University Press.

Morita, T. and H.-C. Lee (1998), IPCC SRES database, version 0.1, emission scenario database prepared for IPCC Special Report on Emissions Scenarios (http:www-cger.nies.go.jp/cgere/db/ipcc.html).

Myhre, G., E.J. Highwood, K.P. Shine and F. Stordal (1998), 'New estimates of radiative forcing due to well mixed greenhouse gases', *Geophysical Research Letters*, **25** (14), 2715–18.

Nakićenović, N. (1996), 'Freeing energy from carbon', *Daedalus*, **125** (3), 95–112.

Nakićenović, N. and R. Swart (eds) (2000), *Emissions Scenarios, Special Report of the Intergovernmental Panel on Climate Change*, Cambridge: Cambridge University Press.

Nakićenović, N., M. Amann and G. Fischer (1997), 'Global energy supply and demand and their environmental effects', 1-6-1 Ohtemachi, Chiyoda-ku, Tokyo 100, Japan, on Agreement No. 96-131 between CRIEPI and IIASA, 28 February 1997.

Nakićenović, N., A. Grübler and A. McDonald (eds) (1998), *Global Energy Perspectives*, Cambridge: Cambridge University Press.

Nakićenović, N., A. Grübler, H. Ishitani, T. Johansson, G. Marland, J.R. Moreira and H.-H. Rogner (1996), 'Energy primer', in R. Watson, M.C. Zinyowera and R. Moss (eds), *Climate Change 1995. Impacts, Adaptations and Mitigation of Climate Change: Scientific Analyses*, Cambridge: Cambridge University Press, pp.75–92.

Odell, P.R. (2000), 'The global energy market in the long term: the continuing dominance of affordable non-renewable resources', *Energy Exploration and Exploitation*, **18** (2&3), 75–92.

Riahi, K. and R.A. Roehrl (2000a), 'Greenhouse gas emissions in a dynamics-as-usual scenario of economic and energy development', *Technological Forecasting and Social Change*, **63** (3), 195–205.

Riahi, K. and R.A. Roehrl (2000b), 'Energy technology strategies for carbon dioxide mitigation and sustainable development', *Environmental Economics and Policy Studies*, **3** (2), 89–123.

Roehrl, R.A. and K. Riahi (2000), 'Technology dynamics and greenhouse gas emissions mitigation: A cost assessment', *Technological Forecasting and Social Change*, **63** (3), 231–61.

Rogner, H.H. (1997), 'An assessment of world hydrocarbon resources', *Annual Review of Energy Environment*, **22**, 217–62.

Schrattenholzer, L. (1999), 'A brief history of the International Energy Workshop', in J. Weyant (ed.), *Energy and Environmental Policy Modeling*, Kluwer's International Series, Dordrecht: Kluwer Academic Publishers, pp.177–85.

Smith, S.J., H.M. Pitcher, M. Wise and T.M.L. Wigley (1998), 'Future sulfur dioxide emissions', mimeo, NCAR, Boulder, CO, USA.

Thaler, R. (1981), 'Some empirical evidence on dynamic inconsistency', *Economics Letters*, **8**, 201–7.

UN (United Nations) (1993a), 'Macroeconomic data systems, MSPA data bank of world development statistics', *MSPA Handbook of World Development Statistics*, MEDS/DTA/1 and 2 June, New York: United Nations.

UN (United Nations) (1993b), 'UN MEDS macroeconomic data systems, *MSPA*

data bank of world development statistics', MEDS/DTA/1 MSPA-BK.93, Long-Term Socioeconomic Perspectives Branch, Department of Economic and Social Information and Policy Analysis, UN, New York.
UN (United Nations) (1996), 'Annual populations 1950–2050: the 1996 revision', UN Population Division, New York, United Nations (data on diskettes).
UN (United Nations) (1998), 'World population projections to 2150', United Nations Department of Economic and Social Affairs Population Division, New York.
Watson, R., M.C. Zinyowera and R. Moss (eds) (1996), *Climate Change 1995: Impacts, Adaptations and Mitigation of Climate Change: Scientific Analyses*, contribution of Working Group II to the Second Assessment Report of the Intergovernmental Panel on Climate Change, Cambridge: Cambridge University Press.
Wigley, T.M.L. and S.C.B. Raper (1997), 'Model for the assessment of greenhouse gas-induced climate change' (MAGICC version 2.3), Climate Research Unit, University of East Anglia, UK.
Wigley, T.M.L., R. Richels and J.A. Edmonds (1996), 'Economic and environmental choices in the stabilization of atmospheric CO_2 concentrations', *Nature*, **379** (January), 240–43.
Wigley, T.M.L., M. Salmon and S.C.B. Raper (1994), 'Model for the assessment of greenhouse gas-induced climate change', version 1.2, Climate Research Unit, University of East Anglia, UK.
WMO/UNEP (World Meteorological Organisation/United Nations Environment Programme) (1998), 'Scientific assessment of ozone depletion: 1998', WMO Global Ozone Research & Monitoring, World Meteorological Organisation (WMO), Geneva.
World Bank (2000), *World Development Indicators 2000*, Washington, DC: World Bank.
World Development Report (1993), World Bank, New York: Oxford University Press.

4. Technology clusters

This chapter presents the concept of technology clusters, a concept that has proved useful for the analysis of long-term energy–economy–environment (E3) scenarios from a policy perspective. Generally speaking, technology clusters are groups of technologies that have at least one important feature in common. We shall define four types of technology clusters, illustrate them with examples, and then proceed to show how the cluster concept can be used to analyse patterns of energy technology evolvement in sustainable-development E3 scenarios.

4.1 DEFINING TECHNOLOGY CLUSTERS

The real-world energy system consists of many thousands of technologies ranging from fuel extraction, refining, energy conversion and conservation to technologies for energy end use. This complexity makes it difficult to assess policies aimed at individual technologies, for example policies to guide the overall energy system towards sustainability. The cluster concept serves the purpose of reducing the complexity of the description of the global energy system. The importance of analysing technologies in such an aggregated way has been widely recognized by other groups; see, for example, Seebregts *et al.* (2000) or Gritsevskii and Nakićenović (2000). Our approach incorporates aspects of both groups when we define some types of technology clusters with respect to technological characteristics prior to – and independently of – modelling, and some types a posteriori from scenario assumptions or results.

We will use the technology cluster concept to analyse the robustness of policies aiming at the promotion of sustainable development (SD) by identifying those principal technology clusters that could accomplish a smooth and efficient transition path from the present energy structure to eventual sustainability, in particular with respect to GHG emissions and their climate impacts. The basic criterion defining the membership of a technology to a cluster is a key feature of that technology, for instance hardware. As we shall see, hardware can refer either to technology components or to an energy-related infrastructure. In both cases, the corresponding clusters are independent of any particular E3 scenario. They can therefore be defined a priori or in absolute terms.

Another kind of possible common feature describes the relation between a technology and the environment in which it operates. Examples of such features are public acceptance and market success. In our analysis, we use these criteria for the definition of two further kinds of clusters.

The public acceptance of future technologies belongs to the general and qualitative assumptions (the 'storyline' as introduced in section 2.3 – pp. 26–35) that stand at the beginning of the definition of an E3 scenario. We assume that the public acceptance of environmentally benign technologies and, accordingly, the willingness to pay an extra price for them, is high in a scenario world in which high priority is given to environmental and sustainability issues. Hence the public acceptance of technologies with high environmental impacts (for example, the use of coal without desulphurization) will be relatively low in SD scenarios. In contrast, the same cluster of technologies might find widespread public acceptance in a business-as-usual scenario (with emphasis on economic growth even at the expense of environmental considerations).

Whereas common public acceptance depends on scenario input assumptions, the market success of a technology can be determined from model results only. Of particular policy relevance are those clusters that are successful in SD scenarios. Our description of how to identify technologies with common market success (section 4.4 below) will therefore use the SD scenarios as a basis. As the latter two kinds of clusters are defined relative to a given scenario we say that they are defined in relative terms or a posteriori.

4.1.1 Absolute Definitions

TP clusters

Common technology components and their manufacturing processes define TP clusters. For example, all fuel cells form a TP cluster. As a consequence, experience gained with fuel cells in the transport sector leads to progress in fuel cell technology in the electricity sector. This definition is significant for the analysis of the long-term development of energy systems in particular when technological progress is included in a formalized way. The reason for this is that any self-consistent scenarios of technological development must take this feature of TP clusters into account because an improvement in one technology is likely to mean technological progress in all the other technologies belonging to the same TP cluster.

Note that one basic technology can be used as a constituent in several compound technologies. This is true of the fuel cell technology already mentioned, but also of the diesel engine, which is used not only in passenger cars, but also in trucks and for stationary applications, for example. The common components that define TP clusters are called 'key technologies'.

IS clusters

Common infrastructure systems define IS clusters. In particular, all technologies with the same fuel input form an IS cluster. For example, all natural gas-related technologies are part of the same IS cluster since they depend on the availability of a natural gas transmission and distribution network. Another example is petrol-driven or diesel cars and room heating technologies using light fuel oil, because they all rely on oil refinery products.

Infrastructures have a service life typically of 50 years or more (Nakićenović *et al.*, 1998). In long-term energy scenarios, IS clusters are therefore important because they introduce significant inertia into the energy system, thus limiting the speed at which fundamental changes in the energy system can occur. An example of such a change would be a hydrogen infrastructure replacing the existing gasoline infrastructure for cars. The common components that define IS clusters are called 'key infrastructures'.

Both energy systems features mentioned in connection with TP and IS clusters – joint technological progress and energy system inertia – play a key role in the so-called 'lock-in' effect. In the case of the inertia introduced by infrastructure systems, lock-in is the result of an existing system's resistance to change. In the case of joint technological progress, lock-in is generated by the increasing competitiveness of progressing technologies, which makes it increasingly difficult for competing technologies to enter the market. Both of these phenomena have to do with increasing returns to scale, and the lock-in is sometimes also referred to as 'path-dependency' of technological change. See Arthur (1990) for a more general analysis of this phenomenon.

4.1.2 Relative (Scenario-dependent) Cluster Definitions

PA clusters

Common public acceptance defines PA clusters. Depending on prevailing value systems, the public acceptance of energy (and indeed all) technology can differ at different times and in different societies. Assumptions regarding future societal preferences are an important part of long-term energy scenarios. PA clusters are therefore scenario-dependent. Two identical PA clusters need not be equally preferred in two scenarios and, moreover, what is a PA cluster in one scenario need not be a PA cluster in another scenario.

A typical example of a PA cluster is one comprising technologies for which environmental and social external costs (for example, agricultural losses from acidification, and social costs of coal mining) are either internalized or low. Other societal values giving rise to the definition of a PA cluster are national security, risk aversion, a preference for centralized over decentralized organization, and lifestyles in general. An obvious special case of a PA cluster is nuclear technology.

PA clusters are defined together with the formulation of the inputs of a scenario. Their implementation in the MESSAGE energy model requires quantifying the 'inconvenience costs' of using electricity rather than wood for cooking, of environmental externalities, or the like (see the appendix).

MS clusters

Common time dynamics of market success define MS clusters. Technologies that, in a particular scenario, increase or decrease their market shares simultaneously therefore form an MS cluster. These clusters are scenario-dependent because they are identified by analysing the time profiles of market shares as given by MESSAGE outputs. MS clusters are related to PA clusters because public acceptance is a determining factor of market succcss, but also TP (for the medium to long term) and IS clusters (for the very long term) play a role in the emergence of MS clusters in different scenarios. One obvious example of MS clusters is crude oil refineries and petrol-driven passenger cars.[1]

4.2 TECHNOLOGY KINSHIP

Over and above introducing the concept of technology clusters to highlight common features of technology, we introduce the concept of 'technology kinship' to describe relationships between technology clusters. In the analysis of long-term energy scenarios, two kinds of such relationships are particularly interesting. One is the strong overlap of clusters of one kind, and the other is the linkage among clusters of different kinds. Such relations among clusters and their technologies determine the dynamic evolution of the energy system and are therefore the focus of interest of policy making. For example, the replacement of clusters with technologies that have high emissions by those that have lower emissions is particularly relevant for SD scenarios.

In order to describe typical patterns of relations between clusters, we give a schematic but comprehensive example of technology clusters in Figure 4.1. The interplay between TP, IS and PA clusters determines the market shares in the scenarios. The IGCC (integrated gasification combined-cycle) technology is used to illustrate the possibility of multiple

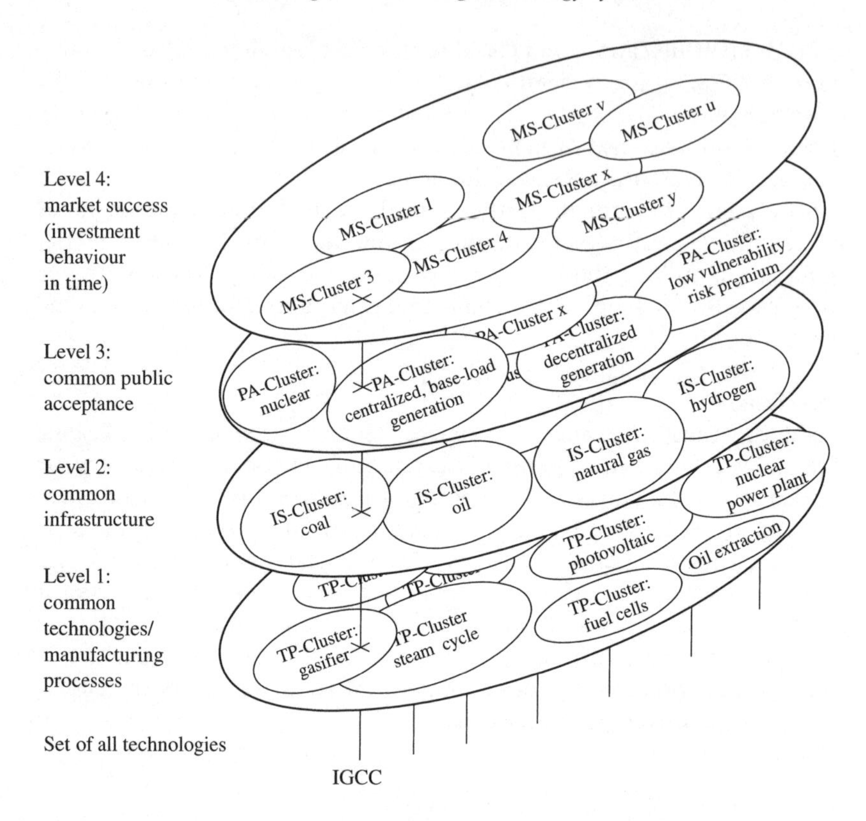

Figure 4.1 *Four kinds of clusters defined by four types of common components*

cluster membership of a technology. Using the IGCC technology as an example, Figure 4.1 illustrates how a technology may, at the same time, belong to clusters on different levels (that is, different kinds of clusters) as well as to two or more clusters on the same level (overlap). IGCC belongs to the coal IS cluster, the PA cluster of centralized base load electricity generation and the MS cluster of high-efficiency fossil electricity generation technologies. It also belongs to the two TP clusters standard 'steam cycle' and 'gasifier'.

With regard to the speed of transformation of the energy system, the PA, TP and IS clusters describe the behaviour of the energy system in the short, medium and long term, respectively. The MS clusters synthesize the temporal behaviour over the total time horizon chosen for the analysis.

4.3 FURTHER EXAMPLES OF TECHNOLOGY CLUSTERS

In this section, we give further illustrations of TP (common technology/process), IS (infrastructure), and PA (public acceptance) clusters. We have chosen them in a way to provide a sound basis for the detailed description of MS (market success), to which we have dedicated a separate section (section 4.4).

4.3.1 TP Clusters and Key Technologies

If the commonality that defines a TP cluster is a technology (and not a technological process), this technology is referred to as a 'key technology' of the corresponding TP cluster. Chip manufacturing would be an example of a key technology that defines a particularly large TP cluster, and the development of solar photovoltaic cells has drawn extensively on the experiences encountered during computer chip production.

The scenario analysis presented later in this book includes TP clusters for nine key technologies in the electricity sector.[2] They are summarized in Table 4.1. The table also identifies 'key fuel infrastructures', which define the membership of each technology in an IS cluster.

As an illustration of the relative importance of TP clusters of power generation technologies, we show, in Figure 4.2, the shares of each TP cluster in global electricity supply in 1990. The figure shows that, in 1990, the single steam cycle dominated electricity production with a market share of approximately 45 per cent. The remainder of the electricity is mainly supplied by three TP clusters: the GC+SC (combined cycle, 20 per cent), R+SC (nuclear reactors including a steam cycle, 17 per cent) and the HT (hydroturbine, 19 per cent) cluster. Wind turbines (WT) also contribute to power production, but their market share is below 1 per cent.

Ranges of projected future contributions of TP clusters to the electricity production in SD scenarios are presented in Figure 4.3. To facilitate the comparison between the power generation structure in SD and that of other scenarios, Figure 4.3 includes 'error' bars indicating maximum and median contributions for other scenarios in cases where the latter exceed the range of the SD scenarios. Light-grey columns indicate the maximum, dark grey columns the median, and black columns the minimum contribution across all sustainable development scenarios.

Table 4.1 Selected aggregate technologies in the electricity generation sector

Abbreviation	Technology Description	Common Component	
		Key technology	Key infrastructure
CoalStdu	Coal power plant, without flue gas desulphurization (FGD) and unabated NO_X emissions	SC	Coal
CoalStda	Coal power plant, 90% FGD, and reduction of 50% of the NO_X emissions	SC	Coal
CoalAdv	Advanced coal power plants; e.g., integrated gasification combined cycle (IGCC)	GF, GC	Coal
FossilFC	Gas- and coal-based fuel cells	FC	Coal, Gas
Oil	Oil power plants	GC or SC	Oil
GasStd	Gas power plant (standard steam cycle)	SC	Gas
GasCC	Gas combined-cycle power plant	GC, SC	Gas
GasReinj	Combined-cycle power plant with no CO_2 emissions (reinjected for enhanced recovery at field), efficiency reduced by 1%	GC, SC	Gas
BioSTC	Biomass power plant (standard steam cycle)	SC	Biomass
Bio_GTC	Biomass gasification power plant	GF, GC	Biomass

Figure 4.3 illustrates that the structure of the electricity sector changes from a present dominance of the SC (steam cycle) cluster, which, by the year 2100, is virtually phased out in all SD scenarios. By that year, the FC (fuel cell) cluster dominates the electricity sector and therefore seems to be the most promising and successful technology option for SD scenarios. On average, fuel cells contribute 38 per cent to the electricity production in

Table 4.1 (continued)

Abbreviation	Technology Description	Common Component	
		Key technology	Key infrastructure
Waste	Waste power plant	SC	Waste collection
Nuc_LC	Conventional nuclear power plant, low costs, low efficiency	R, SC	Uranium
Nuc_HC	Conventional nuclear power plant, high costs, high efficiency	R, SC	Uranium
Nuc&0-Carb	Other advanced zero-carbon technologies (including high temperature and fast breeder reactors)	R, SC or Na/HeC etc	Uranium and others
Hydro	Hydroelectric power plant	HT	Dams
SolarTh	Solar thermal power plant with storage, and solar thermal power plant for H2 production	SC	Solar
SolarPV	Solar photovoltaic power plant (no storage)	PV	Solar
Wind	Wind power plant	WT	Wind
Geotherm	Geothermal power plant	SC	–
H2FC	Electricity from hydrogen fuel cells in the industry and the residential sector, off-peak electricity production via hydrogen-based fuel cells in the transport sector	FC	Mixture
PV-ons	Photovoltaic onsite electricity production	PV	Solar

Notes:
The third column specifies the key technology component; abbreviations: SC (steam cycle), GC (gas cycle), GF (gasifier), FC (fuel cells), WT (wind turbine), PV (photovoltaic modules), R (reactor), HT (hydro turbine), Na/HeC (liquid sodium or helium cycle).

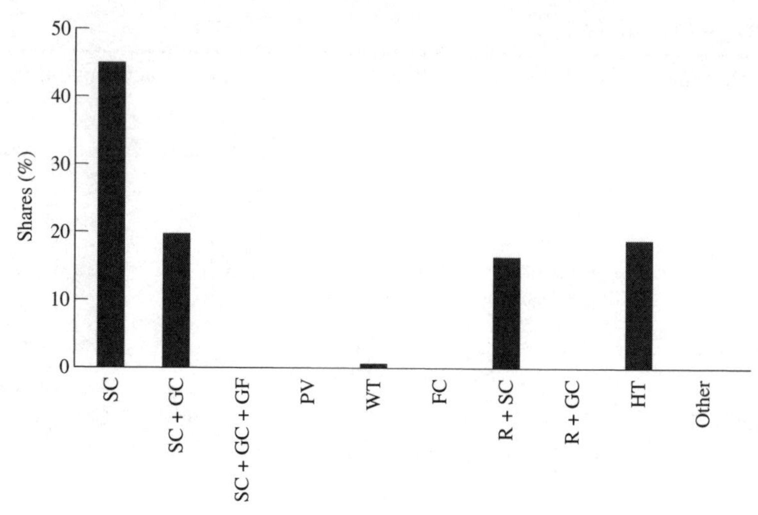

Figure 4.2 Contribution of TP clusters to global power generation in 1990

these scenarios. Their maximum contribution comes close to 50 per cent. Other main contributors to the power generation in 2100 are the PV (photovoltaic modules) and the R+SC (nuclear reactors including a steam cycle) cluster. Minor contributions come from the HT (hydro turbines) and the SC+GC (combined cycle) clusters.

The transition to this largely carbon-free power generation at the end of the 21st century is illustrated by the power generation mix in 2050, which is characterized by evenly distributed market shares of TP clusters. Five dominant TP clusters have median market shares of between 10 and 20 per cent: SC+GC (combined cycle), FC (fuel cells), PV (photovoltaic modules), R+SC (nuclear reactor including a steam cycle) and HT (hydro turbine). The 'error' bars clearly show that the future structure of the power generation system in non-sustainable scenarios may include significantly higher shares of carbon-emitting power generation.

4.3.2 IS Clusters and Common Infrastructures

IS clusters comprise technologies that depend on the availability of a given infrastructure, for example on natural gas extraction, transmission and distribution. Energy infrastructures are not only important for long-term scenarios because they introduce inertia into the energy system, but also because their components are strongly interlinked. Take, for example, a gas extraction facility, which must be closely linked with a gas pipeline or some other network that can deliver the gas to the consumer.

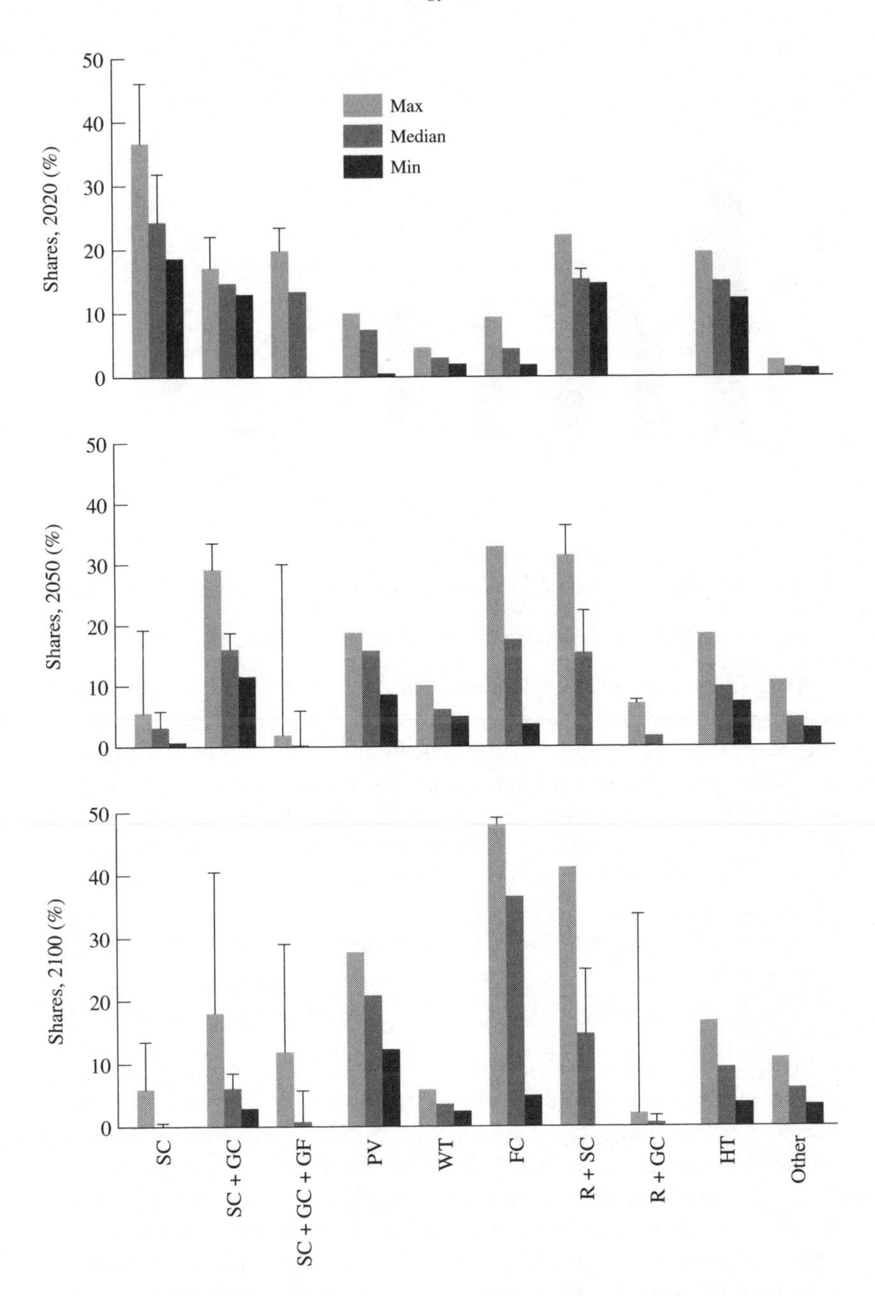

Figure 4.3 Contributions of TP clusters to global power generation in 2020, 2050 and 2100 for sustainable-development scenarios

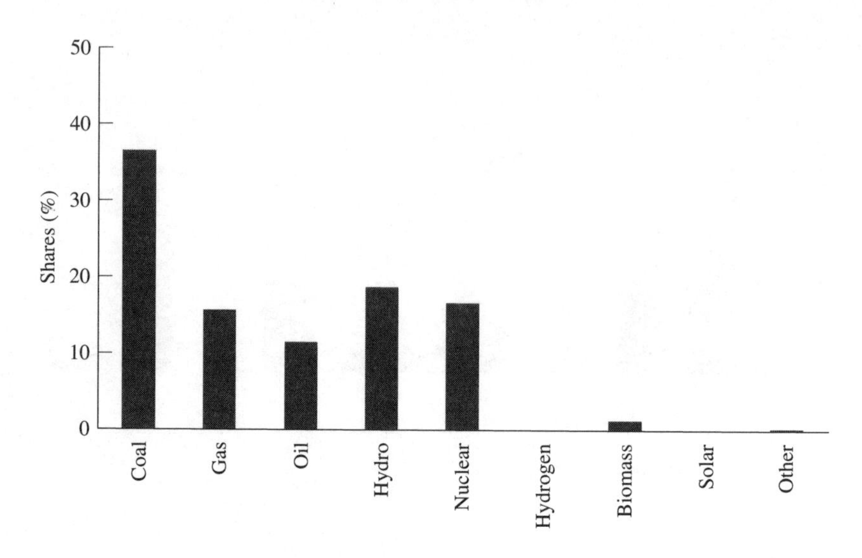

Figure 4.4 Contribution of IS clusters to global power production in 1990

As an illustration of the relative importance of IS clusters of power generation technologies, we show, in Figure 4.4, the shares of each IS cluster in global electricity supply in 1990. The figure shows that global electricity production in 1990 was dominated by the fossil fuels coal, gas and oil, which jointly had a market share of more than 60 per cent. The coal IS cluster is by far the largest contributor of these three with a share of about 36 per cent. The highest shares of non-fossil power come from hydro (19 per cent) and nuclear (17 per cent).

Ranges of future contributions of IS clusters to the global electricity production in SD scenarios are presented in Figure 4.5. The figure shows fuel switches in the electricity sector by depicting results for the years 2020, 2050 and 2100.

Light grey columns give the maximum, dark grey columns the median, and black columns the minimum contribution across all SD scenarios. The 'error' bars indicate maximum and median contributions for other scenarios exceeding the range of the SD scenarios. Most remarkable is the decreasing role, overtime, of the coal and oil IS clusters. Across virtually all SD scenarios their contributions are reduced to insignificant shares by the middle of the 21st century. The only fossil IS cluster that is important for sustainability is the gas IS cluster. Its average market share increases from 16 per cent in 1990 to a peak at about 20 per cent between 2020 and 2050, and decreases to approximately 5 per cent by 2100.

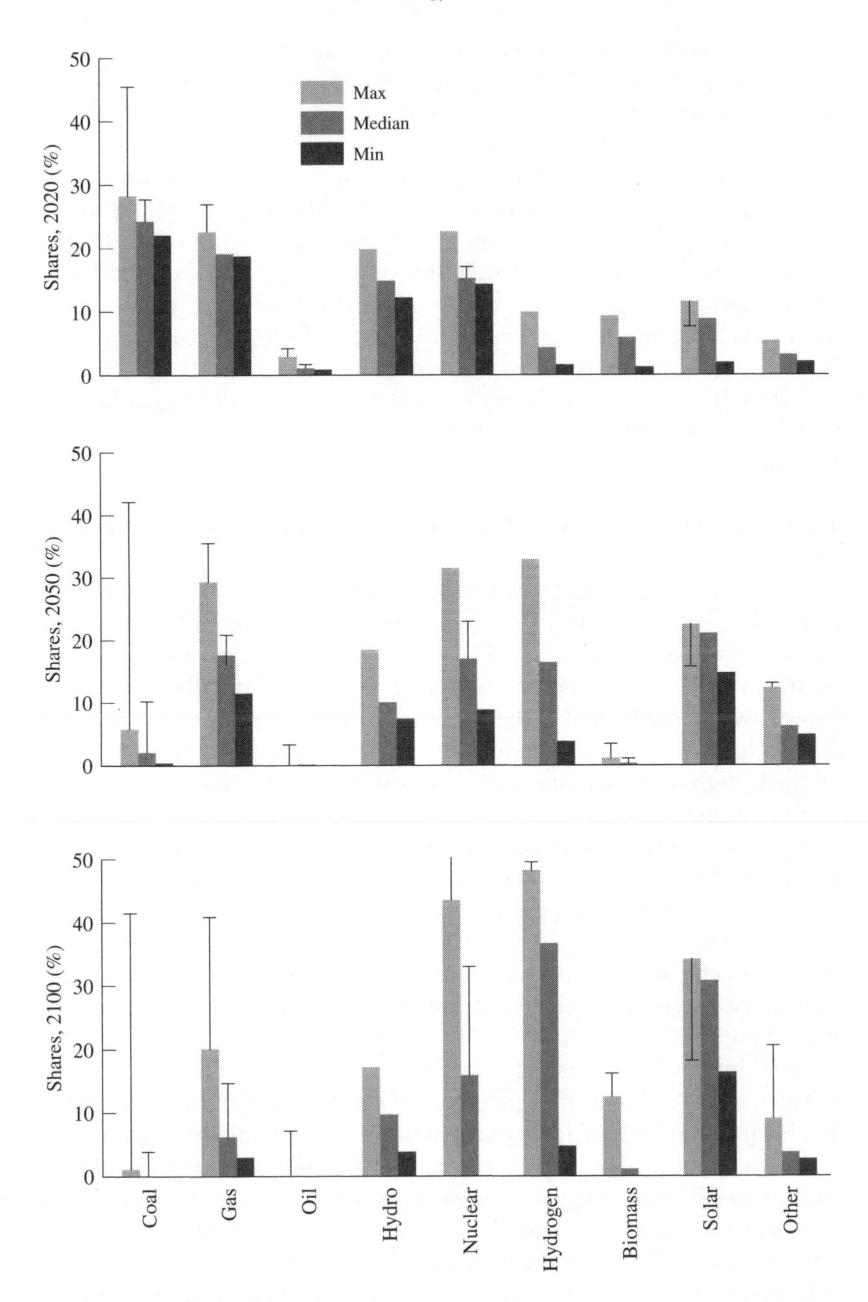

Figure 4.5 *Contribution of IS clusters to the global power generation in 2020, 2050 and 2100 for sustainable-development scenarios*

The non-fossil IS clusters gain significant importance in the medium to long-term future of all SD scenarios (see Figure 4.5). The hydrogen, nuclear and solar IS clusters in particular increase their (average) market shares to approximately 20 per cent as early as 2050, but, in the long run, hydrogen and solar technologies appear to be more compatible with sustainability. Their average market shares increase by 2100 on average to approximately 37 and 31 per cent, respectively (compared to 16 per cent for nuclear). Note that the variability of the projections of the future development of the nuclear IS cluster is very high: its contribution across SD scenarios ranges in 2100 from a total phase-out to market domination (shares of about 45 per cent). Compared to this, the range of solar (between 16 per cent and 31 per cent) is relatively small, suggesting that solar might be a more robust option for achieving eventual sustainability than nuclear.

4.3.3 The Interplay between the Natural Gas and Hydrogen IS Clusters

Like TP clusters, IS clusters can be linked to each other. One key example of a linkage between two IS clusters is the transition between the natural gas and hydrogen IS clusters. Figure 4.6 illustrates how this transition occurs in 13 IIASA scenarios (including scenarios from all three groups) from 1990 to 2100: 1990 is the common origin (20 per cent natural gas, 0 per cent hydrogen). Sustainable-development scenarios are represented by lines, high-impact scenarios by dashed lines and mitigation scenarios by dotted lines.

Figure 4.6 suggests three distinct patterns of the relation between the two IS clusters. One pattern characterizes two high-impact scenarios (B2, A1G) in which the gas share of primary energy supply is high, but where hydrogen supply never exceeds 10 per cent of primary energy. The second kind of relation characterizes the two 'coal' versions of A1 (A1C and A1C-550). In these two scenarios, neither natural gas nor hydrogen plays a major role in the long run.

The third kind of relation characterizes all other scenarios (including the SD scenarios A1T and three members of the B1 family). In those, hydrogen begins to substitute for natural gas around a point where natural gas reaches a market share of 25 per cent. This relation indicates that the natural gas IS cluster plays an important role in paving the way for the emergence of a large hydrogen IS cluster. Also note that SD scenarios are not necessarily those with the lowest natural gas shares.

In particular in the SD scenarios of the B1 family, both natural gas and hydrogen have large shares. The transition patterns are different even among the scenarios with high hydrogen shares towards the end of the

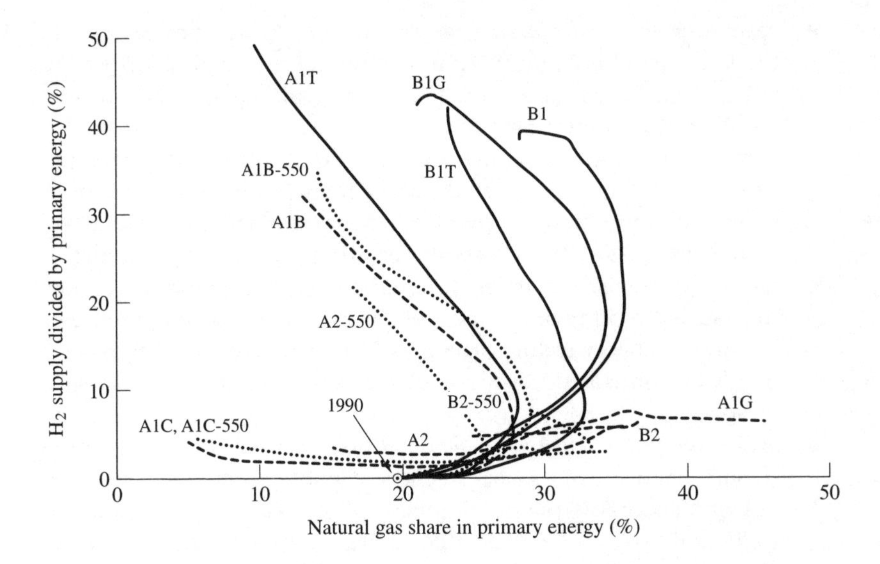

Figure 4.6 Hydrogen supply as percentage of primary energy[3] versus the natural gas share in primary energy, 1990–2100 in 13 IIASA scenarios

century. The transition is particularly smooth in B1 and B1G, where the transition is marked by a maximum use of the natural gas infrastructure. This includes, for example, steam reforming of natural gas and mixing hydrogen with natural gas pipelines (see, for example, Marchetti, 2000).

4.3.4 Public Acceptance Clusters of Energy Technologies

In contrast to TP and IS clusters, public acceptance (PA) clusters are defined by a number of rather intangible commonalties dealing with social values and public preferences. Societal values influencing the public acceptance of energy technologies include the following:

- the internalization of *environmental external costs*. In this respect, all energy technologies with high SO_2 emission intensities form the acidification cluster;
- the internalization of *social external costs*. Social externalities such as risks are involved in deep-coal mining, for instance. Social externalities appear to increase with the increasing general wealth of a society;

- *decentralized v. centralized energy conversion*. Societal preferences of the kind 'small is beautiful' give rise to a cluster that describes low energy technology interconnectedness by grids or pipeline networks. Another important aspect of this PA cluster is that it ranks high in terms of vulnerability of energy supply, for example, to natural catastrophes. As rising shares of the residential sector in total GDP go along with more flexible, decentralized technologies, the prevalence of this PA cluster also reflects structural changes in an economy;
- *security of energy supply*: many countries pay considerable attention to ensuring a minimum degree of national dependency on foreign imports of energy technologies and fuels. The risk associated with a politically motivated disruption of such sources can be valued highly by society;
- *public acceptance of nuclear power*. Looking back in time, the most striking historic example of the effect of public acceptance on a technology's success is the development of nuclear power. In the late 1970s and early 1980s, nuclear power was seen as the most promising technology in the energy sector. It was widely accepted by the public and among energy experts alike. As a consequence, the nuclear industry boomed, and the shares of nuclear electricity generation grew rapidly. Following accidents in nuclear reactors, however, public acceptance decreased in most countries to very low levels, especially in OECD countries;
- *quality of fuels*: another value influencing consumers' choices is the flexibility and convenience of final-energy use. Such criteria explain the use of electricity instead of fuel wood for cooking despite the higher costs of electricity.

Varying assumptions on the future evolvement of societal values can lead to a wide variety of possible future energy technology mixes. PA clusters are therefore an important determinant of long-term energy scenarios, in particular SD scenarios.

Including societal values in the MESSAGE model is accomplished in different ways. High acceptance of environmentally compatible technologies, for example, is modelled by constraints on emissions or the internalization of environmental costs; the shift towards more widely accepted and more flexible energy forms is modelled by so-called 'inconvenience costs'. Inconvenience costs reflect the assumption that consumers are prepared to pay a higher price for more convenient energy services (for example, the preference to heat with electricity instead of using cheaper coal). Finally, societal costs, such as externalities from coal mining, are modelled by cost premiums in addition to the real (physical) costs of the technology.

Table 4.2 gives an overview of electricity generation technologies grouped into three PA clusters of high, low and medium acceptance as identified for SD scenarios. The table shows that, in SD scenarios, it is mainly the renewable energy sources such as solar, wind and geothermal that meet high social and public acceptance. Particularly high is the acceptance of technologies with flexible utilization in combination with renewable fuel consumption, such as photovoltaic-onsite power generation, or power production via small-sized hydrogen fuel cells. It must be remembered, however, that the public acceptance of fuel cells depends on the fuel that is used for the production of hydrogen: the assumed public acceptance of hydrogen fuel cells is only high in the case that the hydrogen comes from renewable sources, and its public acceptance is assumed to decrease with increasing carbon intensity of the hydrogen source.

Five technologies have been identified as meeting low public acceptance in SD scenarios. These are mainly various types of coal and oil technologies, but also include conventional gas technologies with low efficiencies. The cluster labelled 'medium acceptance' can also be interpreted as a 'grey zone' between the clusters with high and low acceptance. This ambiguity expresses the fact that a clear-cut distinction between the PA clusters appears hardly possible. Moreover, the technologies in Table 4.2 can have many different conceivable designs. To reflect this diversity with respect to designs for some technologies, ranges of future public acceptance are reported. For example, 'oil technology' in the PA cluster with low acceptance might be a peak-load diesel engine with *medium* flexibility, but also a conventional oil power plant for base-load application with comparatively *low* flexibility (of utilization).

PA clusters are useful in identifying key technology groups that are compatible with the major societal future concerns assumed to prevail in a scenario. Moreover, they are most helpful in translating the storyline of a scenario into modelling assumptions. However, PA clusters do not give information on which technology group will eventually be the most successful, that is, those technologies that are the key contributors to a given future development. There is a need, therefore, to analyse the success of technologies separately and this is done in the following section, especially emphasizing SD scenarios.

4.4 IDENTIFYING MARKET SUCCESS (MS) CLUSTERS

In the previous section, we first identified the most successful TP and IS clusters in the global power generation of the SD scenarios. We found that

Table 4.2 Four criteria used to classify selected power generation technologies in SD scenarios, according to their assumed future public acceptance. The classification results in three PA (public acceptance) clusters. The ranges for public acceptance reflect different designs of a technology

Technology*	Public acceptance for four criteria			
	Environmental effect	Resource consumption	Risk perceived	Flexibility of utilization
Low PA cluster				
CoalStdu	(very) low	low	low	low
CoalStda	low	low	low	low
CoalAdv	low	medium	low	low
Oil	low	low-medium	low-medium	low-medium
GasStd	medium	low	low	low
Medium PA cluster				
FossilFC	medium	high	low	low
GasCC	medium	medium	medium	medium
GasReinj	high	medium	low	low
BioSTC	high	medium	medium	low
Bio_GTC	high	medium	medium	low
Waste	medium	medium	medium	low
Nuc_LC	medium	medium	low	low
Nuc_HC	medium	medium	low	low
Nuc&0-Carb	medium	medium-high	low	low
Hydro	medium	medium	low	low
High PA cluster				
SolarTh	high	high	medium	low
SolarPV	high	high	medium	low
Wind	high	high	high	low
Geotherm	high	high	medium	medium
H2FC	medium-high**	medium-high**	high	high
PV-ons	high	high	high	high

Notes:
* For a description of the technology abbreviations see Table 4.1.
** The public acceptance of hydrogen fuel cells depends on the fuel input to the hydrogen production. This fuel can be coal or gas, but also renewable energy. We therefore give ranges for its public acceptance.

fuel cells, solar photovoltaic (PV) modules and nuclear reactors (with a steam cycle) were the most successful TP clusters, and we identified hydrogen and solar as the most successful IS clusters, both in SD scenarios. After that we also discussed PA (public acceptability) clusters. On the basis of these three findings on SD scenarios, we now proceed to extend our analysis from selected points in time to the whole time horizon of the 21st century. At the end of this procedure, we will have identified the market success for technology groups, that is, MS clusters.

By the common time dynamics that define market success (MS) clusters, we mean technologies that increase or decrease their market shares simultaneously. This criterion is evaluated for the time trajectories of the market shares of any pair of technologies projected by the MESSAGE model. As we have emphasized, MS clusters can be identified only after scenario results have been obtained. They are therefore scenario-dependent. In the spirit of this book, we are now proceeding to illustrate the identification of MS clusters using the set of SD scenarios.

The market shares of any pair of technologies can evolve in one of three principal possible ways. The two technologies can (a) gain or lose market shares in tandem, (b) substitute one for the other over time, and (c) show neither such dependence across scenarios. To determine which of the three applies, we use the correlation coefficient, which can take values between minus one and plus one. With its help, we quantify the correlation between the market shares of any pair of technologies for a given group of two or more scenarios.

A positive correlation ($R^2 > 0$) indicates cluster-like dynamics (with $R^2 = 1$ expressing perfect clustering) in all scenarios in the group analysed; that is, a situation in which market shares of the two technologies evolve exactly in parallel. A negative correlation ($R^2 < 0$) indicates technology substitution (with $R^2 = -1$ expressing perfect substitution) in all scenarios in a given group. An insignificant correlation ($R^2 \sim 0$) indicates a 'bridging' technology and signifies different behaviour in the various scenarios of the underlying group. These possibilities are not exhaustive, but they are all that is needed for the definition of MS clusters. We illustrate these three possibilities in Figure 4.7.

With this classification, we now proceed to define a single *key technology cluster* as being the most characteristic of the power generation sector in SD scenarios.

4.4.1 Defining the Key Technology Cluster

In this section, we introduce the concept 'key technology cluster' and apply it to identify the 'key market success' cluster of SD scenarios, illustrating it with an example that is most relevant for sustainability.

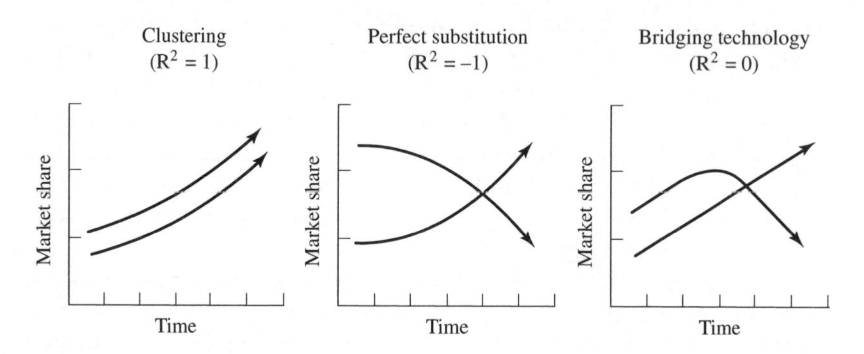

Figure 4.7 Schematic illustration of three different types of relation between two technologies

The analysis of TP and IS clusters presented in section 4.3 provided important insights into the eventual contribution of TP and IS clusters to global energy supply in SD scenarios. There our analysis was limited to selected points in time. We shall now expand our static analysis by looking into the time dynamics of market success for distinct technologies in the electricity sector.

As a point of departure, we took the two sets of TP and IS clusters with the most successful technologies as identified in section 4.3. The two sets include the most promising technologies with respect to their technology components (nuclear reactors with steam cycle, fuel cells and solar photovoltaic modules) and fuel consumption (hydrogen and solar energy) respectively. We then calculated, for each pair of technologies that belong to both types of most successful clusters, their correlation coefficients to determine those with high positive correlation; that is, technologies that develop in parallel over the time horizon. Those technology clusters that have rising market shares are interpreted as accomplishing the structural shifts required for sustainability because, first, they are most successful across SD scenarios from both the TP and the IS point of view and, second, they develop in parallel over the whole time horizon.

Figure 4.8 illustrates the identification of the key MS cluster as the merger of the most successful TS and IS cluster technologies in the power sector. It identifies hydrogen fuel cells and solar photovoltaic conversion as playing a key role across all SD scenarios. The key MS cluster of SD scenarios in the power sector consists of two technology types only: the hydrogen fuel cell and the solar photovoltaic technology. The R^2 of this technology pair is 0.81, illustrating the almost parallel and persistent

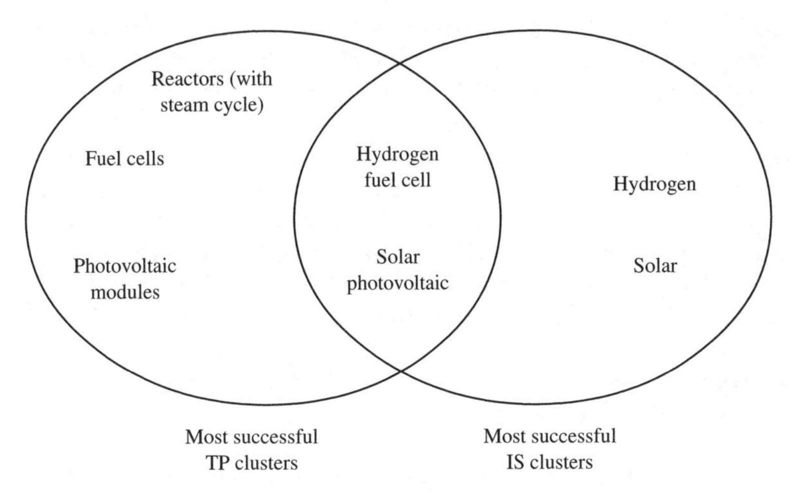

Figure 4.8 The key MS cluster of SD scenarios

diffusion of these technologies across all SD scenarios. This remarkable agreement across scenarios, which is the result of analysing them from multiple viewpoints (TP, IS and MS cluster analysis), highly recommends these technologies for target technology policies in the future. According to this analysis, they seem to be the most promising option for directing the present energy structure towards sustainability.

Calculating the correlation coefficients between other technologies (which need not belong to the most successful TP and IS clusters) and those from the key MS cluster allows us to identify technology options that also appear promising, albeit to a lesser extent. To this group belong advanced nuclear power (with correlation coefficients between 0.58 and 0.68), solar thermal electricity conversion (0.77 to 0.87) and wind (0.52 to 0.81). Note that these technologies are important across all SD scenarios, but their diffusion into the future market is less pervasive and successful than for the key MS cluster technologies.

The group of technologies with particularly low R^2s compared to the key MS cluster technologies can be identified as the least relevant for sustainability. In most scenarios of the set, these technologies are replaced by the emerging group of sustainable technologies of the key MS cluster. The replaced technologies are predominantly coal and oil-fired technologies, gas technologies with comparatively low efficiencies, biomass technologies without gasifiers, present-day nuclear technologies (heavy and light water reactors), and waste incineration power generation technologies.

Finally, the technology group with correlation coefficients close to zero (no correlation with key MS cluster technologies) is also relevant for the understanding of SD scenarios. This group mainly consists of technologies that are only successful in the near future, thus forming a bridge to an energy system that is eventually dominated by the key MS cluster. Typical members of this group are the gas combined-cycle technology (with correlation coefficients between −0.36 and 0.13) and fuel cells fuelled by fossil energy, predominantly gas (−0.26 to 0.14). These technologies are interpreted as having the highest prospects of initiating those structural changes that in the long run achieve a smooth transition to sustainability.

4.4.2 Frequency Distributions of Technology Market Shares for a Given Point in Time

In the previous section we discussed market success clusters in terms of market share changes over a given time horizon. This was done with the simple statistical method of calculating correlations for time trajectories of market shares for the various technologies. Essentially, this method is good for identifying similarities in time dynamics and patterns, assigning each time interval the same 'weight' in the analysis. An alternative way of using time profiles of market shares of technologies for analysis is to pick out snapshots of market shares of a given technology at a given point in time. When this is done for many scenarios for the same fixed point in time, we can portray the complete frequency distribution of technology market shares. Analysing these frequency distributions permits identifying most likely or 'robust' outcomes under a range of possible assumptions.

This type of analysis also expands the analysis presented in an earlier part of this chapter (section 4.3) where we reported medians and ranges derived from frequency distributions of market shares of many technologies at a time. However, medians and ranges are only rough statistical indicators of a data set and do not reveal much about the actual distributions. To guide our scenario results further in the direction of entire distributions, we now present, in Figure 4.9, the entire frequency distributions of all natural gas technologies. In our opinion, this is one of the the most important groups of technologies, which may serve as the key bridge between today's energy systems and a sustainable one.[4]

Our first general observation is that the resulting histograms do not resemble Gaussian (that is, normal) distributions – with the possible exception of the first few decades of the 21st century. Later the frequency distributions have two, three or more separate peaks (modes), indicating that market shares near these peaks of the distribution are more frequently observed than others. We now explain this multi-peak behaviour with the help of cluster analysis.

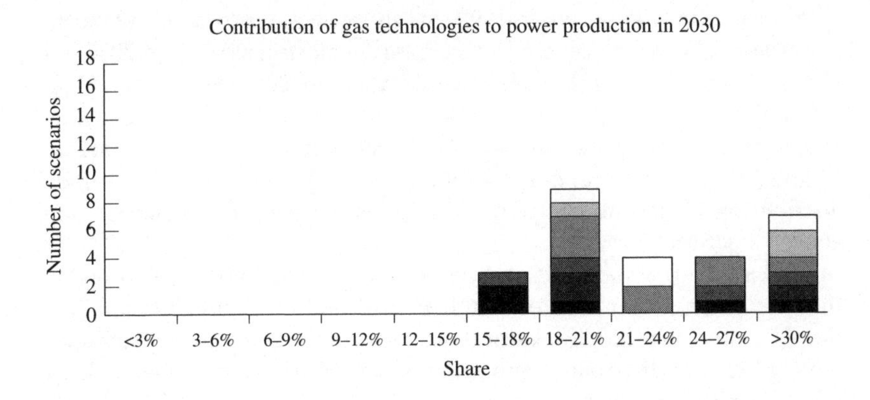

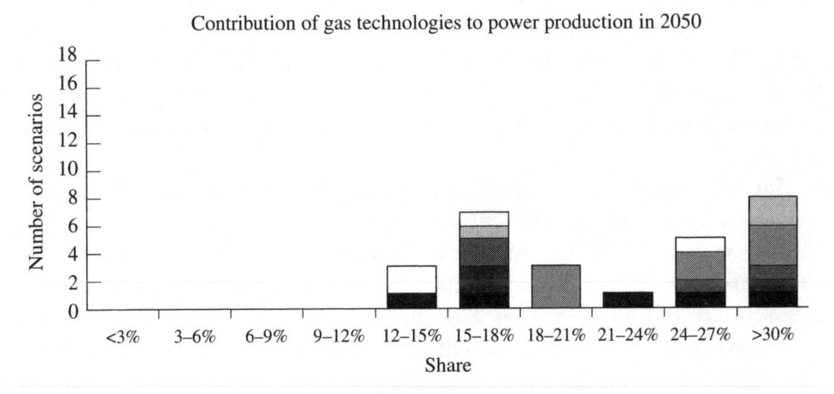

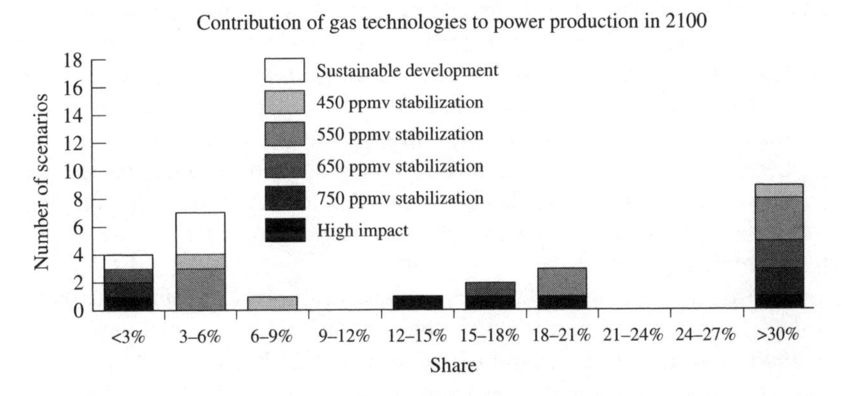

Figure 4.9 *Frequency distribution of market shares of the IS cluster of gas technologies in total world power generation in 2030, 2050 and 2100*

Starting from 14 per cent in 1990, the share of the natural gas IS cluster in the world electricity generation increases in all scenarios up to 2030, but already shows two gentle peaks in that year. In 2050, these peaks have already become rather distinct. In the second half of the 21st century, the higher of these two peaks (around the 20 per cent market share) bifurcates again so that the natural gas market shares in 2100 exhibit a three-modal distribution, clustering around market shares of 3 per cent, 17 per cent, and above 30 per cent.

Looking at the stabilization scenarios only (see Figure 4.9) suggests that there is no clear relationship between the long-term CO_2 stabilization level and the share of natural gas technologies in power generation: their shares are distributed in no regular way. The same applies to high-impact scenarios (black). Only in the SD scenarios (white boxes) are the shares of natural gas technologies found near the low end of the scale. This reflects a transition to an energy system in which hydrogen produced from renewable energy sources plays a major role because renewable sources are preferred to using up the fossil resource base in a sustainable world.

The fact that climate control policies do not appear to have a clear impact on long-term natural gas use suggests that investing in natural gas technology would be a robust strategic decision, largely independent of future policies. Even in SD scenarios, we see a significant potential for growth in world gas use at least until 2040. Whether the world will start to phase out gas use thereafter will be determined mainly by two factors. The first factor is the technological progress (in terms of cost and efficiency improvements) of the gas combined-cycle and natural gas fuel cell technologies. This factor explains the first bifurcation by 2050 (Figure 4.9). The second factor, international gas infrastructure (pipelines, LNG – liquefied nature gas – ports) policies (Klaassen *et al.*, 2000), combined with productivity improvements in natural gas extraction, explains the second bifurcation. Using our cluster terminology, we have shown how the interplay between IS and TP clusters creates two bifurcations in the frequency distribution of market shares of natural gas technologies.

Looking now more specifically at the gas combined-cycle (GasCC) technology (Figure 4.10), we find a rapid increase in its shares, from almost zero in 1990 to around 14 per cent in 2030. Between 2030 and 2050, the distribution widens, while the median share hardly moves. In the second half of the 21st century, GasCC is nearly phased out in the SD scenarios (where we observe a shift to decentralized production of hydrogen with ensuing electricity production) and in the mitigation scenarios (in which energy supply shifts to high-efficiency fuel cells and/or other fuels). In contrast, most

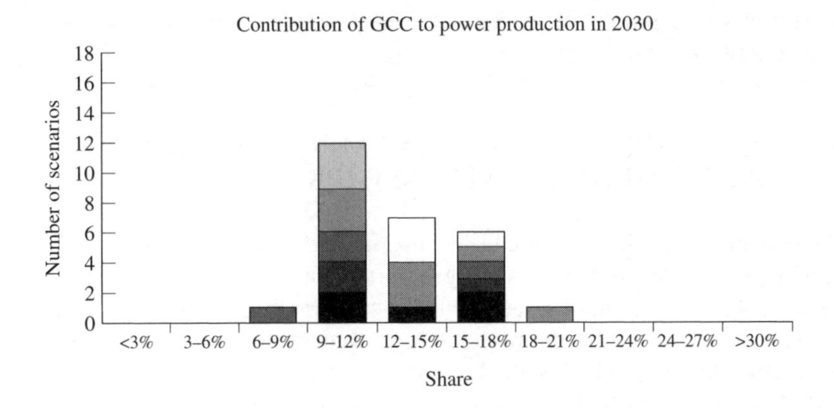

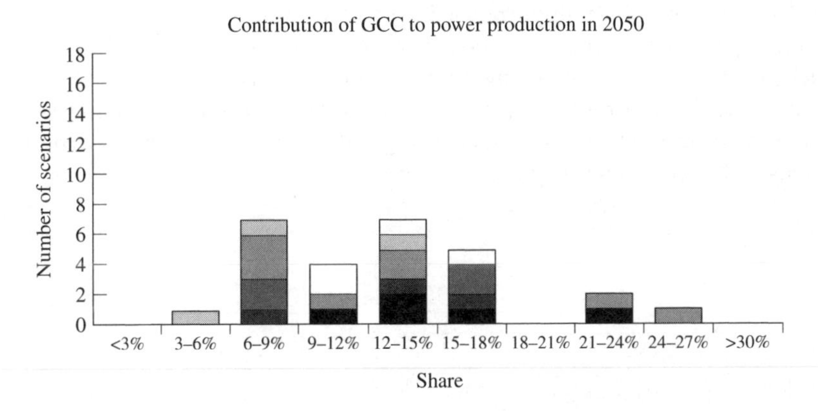

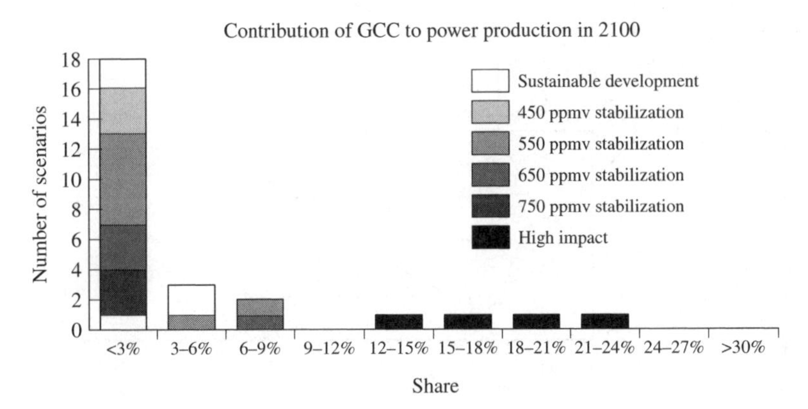

Figure 4.10 Frequency distributions of market share of gas combined-cycle technologies in total world power generation in 2030, 2050 and 2100

high-impact scenarios show GasCC market shares between 12 and 24 per cent as late as 2100.

4.5 A COMPREHENSIVE EXAMPLE OF CLUSTERS AND THEIR RELATIONSHIPS

We conclude this chapter on cluster analysis with a schematic illustration (Figure 4.11) of technology clustering in the electricity sector that is common to most of our SD scenarios, such as the B1 scenario family described earlier.

Technologies with common technology components are connected by lines, thus denoting TP clusters. Dotted lines denote technologies that are only 'weakly' connected, in the sense that the common technology components are of minor importance in terms of total costs.[5] The circled numbers indicate similar technologies in the sense that they have the same fuel input, that is they belong to the same IS cluster. The horizontal arrangement of the boxes illustrates the dynamic evolution of MS (market success) clusters. The vertical axis arranges the boxes by declining carbon intensities of the technologies in the box.

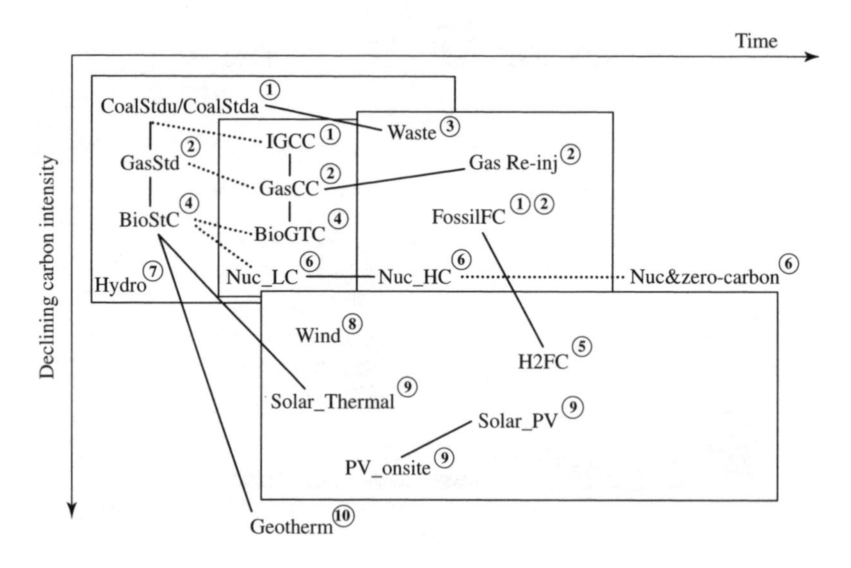

Note: Abbreviations are explained in Table 4.1.

Figure 4.11 Technological clusters in an SD scenario

The figure shows successive substitutions of MS clusters over time. At present, the MS cluster with conventional technologies (CoalStdu/CoalStda, GasStd, BioStC and so on) dominates. In the long run, an MS cluster consisting of wind, solar thermal (SolarTh), solar photovoltaic (SolarPV) and hydrogen fuel cell technologies emerges.[6]

In order to emphasize the most important features of this analysis, the PA clusters have not been highlighted separately. However, in an SD scenario such as the one described in the figure, in which a high degree of environmental protection is assumed, the carbon intensity of the technologies (vertical axis) may be taken as a proxy for the technology's public acceptance. The upper part of Figure 4.11 shows technologies with high carbon intensity, which therefore belong to PA clusters with low public acceptance. By the same token, the lower part of the figure shows PA clusters with high public acceptance.

NOTES

1. Refineries and petrol-driven cars belong not only to the same MS cluster but also to the same IS Cluster.
2. Seebregts *et al.* (2000) also use clusters (TP clusters in our definition) in their energy model. They identified 20 clusters for the energy system as a whole and they use five key technologies (photovoltaic modules, wind turbines, fuel cells, gasifiers, gas turbines) for the five clusters they include in the electricity sector. Similarly, Gritsevskii and Nakićenović (2000) use ten technology clusters in their energy model.
3. Note that hydrogen is not primary energy. Still, the ratio of hydrogen supply (measured in secondary energy units) to primary energy appears to be a good measure of the importance of the hydrogen IS cluster in the scenario.
4. This subsection is the only place in this book where we include results of scenarios with CO_2 stabilization levels other than 550 ppmv, mainly for the reason of increasing the sample size for the histograms.
5. These may nonetheless denote important technology transitions from an overall system's perspective.
6. For a description of the technology abbreviations see Table 4.1.

REFERENCES

Arthur, W.B. (1990), 'Positive feedbacks in the economy', *Scientific American*, **262** (February), 92–9.

Gritsevskii, A. and N. Nakićenović (2000), 'Modeling uncertainty of induced technological change', Energy Policy, **28**, 907–21.

Gritsevskii, A. (1996), 'Scenario generator', internal report, International Institute for Applied Systems Analysis, Laxenburg.

Klaassen, G., K. Riahi and R. Roehrl (2000), 'Global energy scenarios, gas transmission and the environment in Asia', paper prepared for the Northeast Asian Gas and Pipeline Forum, hosted by the Energy Systems Institute, Irkutsk, 17–19 September.

Marchetti, C. (2000), 'On decarbonization: historically and perspectively', prepared for HYFORUM 2000, Munich, September 2000.
Nakićenović, N., A. Grübler and A. McDonald (eds) (1998), *Global Energy Perspectives*, Cambridge: Cambridge University Press.
Seebregts, A., T. Kram, G.J. Schaeffer and A. Bos (2000), 'Endogenous learning and technology clustering: analysis with MARKAL model of the Western European energy system', *International Journal of Global Energy Issues*, **14** (1–4), 289–319.

5. A sustainable-development scenario in detail

In this chapter, we describe one sustainable-development (SD) scenario in detail. The purpose of this description is to provide the readers with a basis for a more detailed judgmental assessment of the scenario and its determining assumptions. We will present these from a policy perspective; that is, we emphasize the description of those parameters and variables that appear particularly interesting for policy making. These include overall and per capita economic growth as well as technological progress. Of these, technological progress is the one that is the focus of our modelling and assumed to be influenced by policy making. In our interpretation, policy making aimed at the support of appropriate technologies can help pave the way for sustainable development. As a means of emphasizing salient features of the SD scenario described in detail in this chapter, we contrast them with assumptions that lead to a non-sustainable scenario in a similar 'world'; that is, in a world in which the key boundary conditions are by and large the same and just the policies are different.

To emphasize the policy orientation, we illustrate how different policies contribute to the different developments of the scenarios. The assumed policies differ with respect to the direction of technological innovations. Two different sets of assumptions regarding different policy options lead once to an 'oil and gas'-rich future (OG), and once to a 'post-fossil fuel' future (PF).[1] The latter turns out to be an SD scenario. To express this policy aspect, we also use the term 'strategy' synonymously with 'scenario', in particular in those places where we want to emphasize the policy relevance of a particular point.

'Post-fossil' means that technological progress is concentrated on conversion technologies fuelled mainly by renewable energy, on technologies that produce or utilize synthetic fuels including hydrogen, as well as on efficiency improvements of end-use technologies. The 'oil and gas'-rich strategy is a future in which technological change is concentrated on the oil and natural gas sectors, including extraction and refinery technologies. Accordingly, technological progress in this scenario is reflected by cost reductions in unconventional oil and gas extraction and conversion technology and substantial improvements and extensions of the present pipeline grids, among others.

It is important to note that these characterizations are parts of the 'storylines' of the two scenarios and not the results of formal modelling. This is important for policy making because it means that the scenarios include steeper cost reductions for those technologies that are assumed to be the target of R&D and other technology policies. Based on these cost trajectories, the scenario suggests a specific (cost-optimal) development of the global energy supply mix, consistent with the qualitative assumptions in the storyline. Although the postulated cost reductions cannot be predicted with certainty, the scenarios describe possible consequences of policy making, and not even to consider the strategies would in our opinion make the scenarios unlikely, if not impossible. This is particularly true for the SD scenario.

The following is a much more detailed version of the scenario descriptions of Chapter 3. Although it reflects the structure of the modelling tools used to generate the scenarios, we think that, for a general understanding of the flow of our arguments, no particular detailed knowledge of the tools is required. Nonetheless, for the readers also interested in the methods used, we describe mainly MESSAGE, but also MESSAGE-MACRO and the Scenario Generator in the appendix to this book.

The structure of this chapter is as follows. First, the assumptions made for both scenarios are described and then the results are presented and discussed from a policy perspective. In the third section, we present an order-of-magnitude estimation of the R&D efforts required to achieve the technological progress assumed in the SD PF scenario.

We first present the assumptions on demographic and economic development which are common to both scenarios and go on to present the assumptions that describe the difference between the two technological strategies. Numerical assumptions for MESSAGE's 11 world regions are summarized and presented in aggregated form for four 'macro' world regions.[2]

5.1 NUMERICAL ASSUMPTIONS FOR THE TWO SCENARIOS

5.1.1 Basic Scenario Assumptions

Many crucial assumptions about long-term E3 scenarios are qualitative. Most prominently, these include general assumptions about economic growth. But, even after quantification, numerical values of economic growth and economic output do not enter MESSAGE directly. Still they are very important indirect determinants of MESSAGE inputs because, together with assumptions on the development of energy intensity of GDP, they determine energy demands, which are direct inputs to MESSAGE. The overall economic and demographic assumptions are made common to both scenarios to facilitate their overall comparability.

5.1.2 Economic and Population Growth

Population

Of the existing scenarios of population growth in the 21st century, we chose one consistent with the empirical observation that higher affluence goes together with families with fewer children (Barro, 1997). For the OG and PF scenarios, the low variant of the projections by Lutz *et al.* (1996, 1997) was selected. For the numerical values of the projected population, see Table 5.1 and Figure 5.1.

Table 5.1 Population (millions) assumptions, values for four world regions and average annual growth rates, OG and PF scenarios*

Region	1990 (actual)	2020	2040	2060	2080	2100
OECD	859	1007 (0.5%)	1069 (0.3%)	1084 (0.1%)	1098 (0.1%)	1110 (0.1%)
Asia	2798	3937 (0.9%)	4238 (0.4%)	4085 (−0.2%)	3589 (−0.6%)	2882 (−1.1%)
REFS	413	433 (0.2%)	433 (0.0%)	409 (−0.3%)	374 (−0.4%)	339 (−0.5%)
ROW	1192	2241 (1.9%)	2791 (1.1%)	3089 (0.5%)	3064 (0.0%)	2727 (−0.6%)
World total	5262	7618 (1.1%)	8531 (0.6%)	8667 (0.1%)	8125 (−0.3%)	7058 (−0.7%)

Note: *Over two decades ending in the given year.

It is assumed that after peaking at 8.7 billion in the middle of the 21st century, world population will decline to 7.1 billion in the year 2100. The assumptions of below-replacement fertility levels and increasing life expectancy lead to significant population aging, which in the long term affects all world regions.

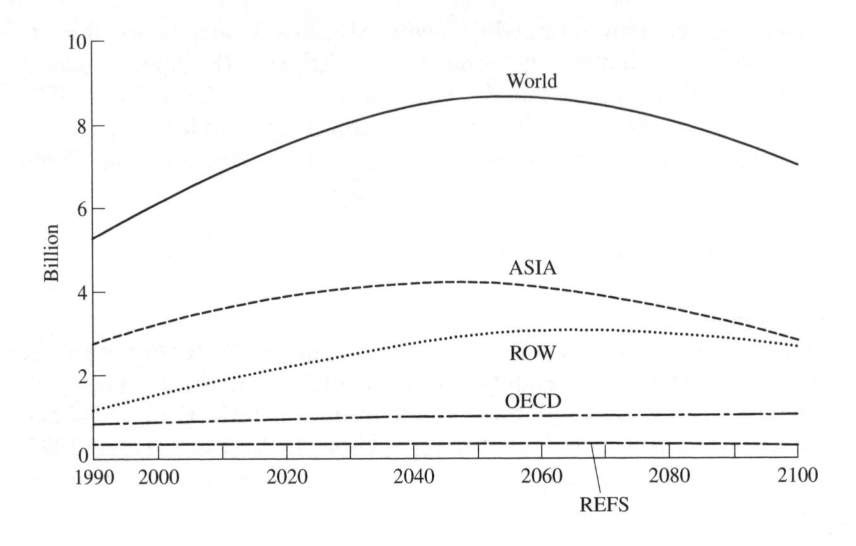

Figure 5.1 Global and world regional population developments for 1990–2100, assumptions used in the OG and PF scenarios

Economic growth

The main theme of this book is to portray technological paths that can lead to a sustainable development of the global E3 (energy–economy–environment) system and to discuss policy aspects of supporting such a development. As is well known, policy making in general is made easier when economic growth is fast. Add to this the fact that high economic growth also means a faster introduction of new (and the replacement of old) energy-converting equipment, and it is found most plausible to assume high economic growth as the background for describing successful policy action for sustainable development; this is what we have therefore chosen for the 'storyline' of the SD scenario. In contrast to the often-quoted slogan, 'small is beautiful', as a characterization of sustainable development,[3] our SD scenario also suggests that sustainable development and high economic growth are compatible.

Overall economic output can be thought of as the product of per capita GDP and population. We begin with a documentation of our assumptions with respect to per capita GDP for the four aggregated world regions in Table 5.2. The graphical illustration of these assumptions is given in Figure 5.2.

Per capita incomes in industrialized countries (OECD plus Reforming Economies) are assumed to increase approximately to 109 000 (constant 1990 US$) and to 70 000 (constant 1990 US$) in other countries (Asia plus

Table 5.2 GDP per capita (thousand 1990 US$ per year) assumptions, absolute values for four world regions and average annual growth rates, OG and PF scenarios*

Region	1990 (actual)	2020	2040	2060	2080	2100
OECD	19.1	31.4 (1.7%)	43.8 (1.7%)	60.6 (1.6%)	82.7 (1.6%)	112.0 (1.5%)
Asia	0.5	3.4 (4.7%)	10.6 (7.5%)	23.5 (3.0%)	43.3 (2.5%)	75.7 (2.2%)
REFS	2.7	4.8 (7.1%)	21.7 (5.6%)	39.6 (4.0%)	65.2 (3.1%)	101.5 (2.8%)
ROW	1.6	4.4 (4.5%)	12.3 (5.2%)	24.8 (3.5%)	40.0 (2.4%)	63.5 (2.3%)
World total	4.0	7.5 (2.7%)	15.9 (3.8%)	29.3 (3.1%)	48.4 (2.5%)	78.0 (2.4%)

Note: *Over two decades ending in the given year.

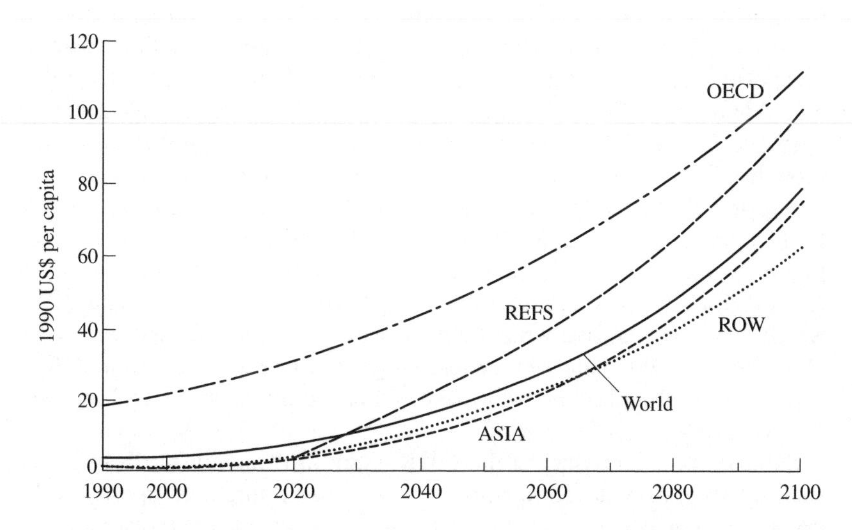

Figure 5.2 Global and world regional GDP per capita developments for 1990–2100, assumptions used in the OG and PF scenarios

Table 5.3 Summary of the assumed GDP (trillion (10^{12}) 1990 US$) assumptions, values for four world regions and average annual growth rates, OG and PF scenarios*

Region	1990 (actual)	2020	2040	2060	2080	2100
OECD	16.4	31.6 (2.1%)	46.8 (2.0%)	65.7 (1.7%)	90.8 (1.6%)	124.3 (1.6%)
Asia	1.5	13.5 (8.0%)	44.9 (6.0%)	95.8 (3.8%)	155.5 (2.4%)	218.2 (1.7%)
REFS	1.1	2.1 (4.8%)	9.4 (7.5%)	16.2 (2.7%)	24.4 (2.0%)	34.4 (1.7%)
ROW	1.9	9.8 (6.4%)	34.3 (6.3%)	76.5 (4.0%)	122.5 (2.4%)	173.1 (1.7%)
World total	20.9	57.0 (3.8%)	135.4 (4.3%)	254.1 (3.1%)	393.2 (2.2%)	550.0 (1.7%)

Note: *Over two decades ending in the given year.

Rest of the World). The per capita income ratio between the developing and the industrialized regions is thus assumed to be 1:1.6 in 2100, compared to a ratio of 1:16 in 1990. GDP per capita, together with population assumptions (Table 5.1), gives GDP development in the four world regions (Table 5.3). Again, a graphical presentation of the result is given in Figure 5.3.

In the OG and PF scenarios, the assumed massive global economic growth includes the hypothesis that productivity in developing countries can approach the levels of industrialized countries following a path similar to the postwar growth in Japan and South Korea as well as the recent economic development of China. Overall, the global economy is projected to expand at an average annual rate of 3 per cent between 1990 and 2100, which is roughly in line with the historical trend over the last 100 years (Maddison, 1995). This average annual growth rate amounts to a 26-fold expansion of global economic output, which would reach 550 trillion 1990 US$ by 2100.

Some readers may consider this GDP assumption unrealistically high, yet such a setting provides for a consistent basis to explore high rates of technological change, which we assume to favour massive R&D investment.

Market exchange rates and purchasing power

Note that, throughout this book, we measure GDP using market exchange rates. Traditionally, market exchange rates are used in international

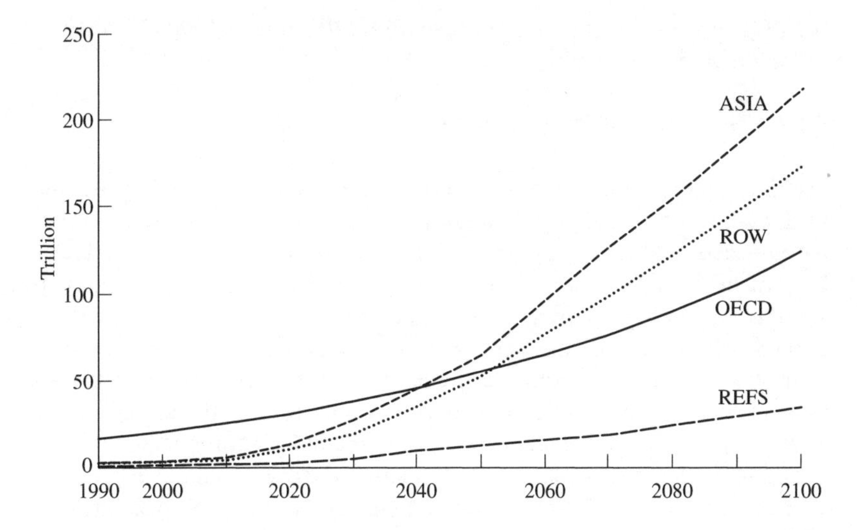

Figure 5.3 Global and world regional GDP developments; projections for 1990–2100, OG and PF scenarios

comparisons, in which economic indicators are converted from local currencies into a common currency, such as US dollars. Only in theory do exchange rates adjust so that the local currency prices of a group of identical goods and services represent equivalent value in every economy. In practice, such adjustment processes can be slow and lag far behind changing economic circumstances. Policies such as currency controls may further distort the accuracy of market-based rates. Moreover, many goods and services are not traded internationally. For these, international price adjustments might be particularly slow (Nakićenović and Swart, 2000).

In short, this means that the purchasing power of one US dollar may be different in different countries. One attempt to quantify these differences was undertaken by the International Comparison Project, which compared prices for several hundred goods and services in a large number of countries. On the basis of this comparison, the relative values of local currencies are adjusted to reflect the purchasing power parity (PPP) of currencies (UNDP, 1993). The biggest differences between monetary values that are expressed in market exchange rates and those that are corrected using purchasing power parity have been observed for developing countries.

A systematic and consistent use of the PPP concept requires making some assumptions, in particular with respect to the speed of adjustment of the purchasing power in different economies. A detailed treatment of this issue is beyond the scope of this book.[4] We therefore just note that using the PPP concept leads to a higher global GDP value in 1990 (25.7 instead of 20.9 trillion 1990 US$).

5.1.3 Technological Progress and Energy Demand

Energy demand inputs for MESSAGE are the result of combining assumptions on economic and population growth with assumptions on technology development. More precisely, a separate model called the 'Scenario Generator' (SG) is used to transform qualitative assumptions about the speed of technological progress into quantitative assumptions on the development of the overall final-energy intensity of GDP. The SG combines extensive historical economic and energy data with empirically estimated equations of trends in order to determine future structural change (Gritsevskii, 1996). For each world region, the SG generates future paths of final-energy demand consistent with historical dynamics and specific scenario characteristics. The resulting final-energy demands are disaggregated into seven energy demand sectors.[5]

We shall divide the description of technological progress assumed for the OG and PF scenarios into three areas: resource extraction technologies, electricity generation and synthetic fuel technologies, and end-use technologies. For electricity generation and synthetic fuel technologies, the assumptions on technological progress are specified mainly as the speed of technology's (unit investment) cost reduction as well as the earliest availability (at the assumed cost) of new technologies. In addition, the availability of renewable resources is limited by a maximum annual energy potential in each region. The assumptions on technological progress on the end-use side are given as energy intensity reductions over time.

In contrast to using common assumptions for economic and population growth for PF and OG, we use different assumptions for technological progress for the two scenarios. Underlying different paths of technological progress is the assumption that strategic choices in the field of technology policy, in particular R&D policy, can make a major difference to the performance of the energy system. In MESSAGE, such qualitative assumptions on the strategies are the basis for the numerical assumptions of technological change of individual technologies. We want to emphasize that, at this point, we have not endogenized (and thereby optimized) this interplay,[6] but we will use other parts of IIASA-ECS work to give us some very rough quantitative results on the costs of such R&D strategies (see Section 5.3).

Technological progress of extraction technologies

For primary-energy extraction over the course of a century, technological progress is relevant in two respects. One is the total availability of a resource in the form of recoverable reserves. As we have explained above (see the discussion of the McKelvey diagram in Figure 2.3), this number reflects not only a limit set by nature, but also the result of technological advancement of exploration. The second scenario-relevant aspect of primary-energy reserves is economic recoverability over time, which is a function of market prices and advancement of extraction technology.

Both of these aspects are important for the formulation of inputs into MESSAGE, which on the primary-energy side consist of two kinds of parameters. The first kind describes resource recoverability in terms of total availability over the time horizon (stocks) and the second annual extraction limits (flows) of primary energy. Each primary-energy carrier is divided into cost categories to reflect increasing extraction costs as a consequence of increasing technical complexity and scarcity. Conceptually, extraction costs include the cost of exploration and royalties, among other things. On the basis of the extraction costs in each category, the model decides whether it is economical to tap a given resource category, thus converting it from a resource into a reserve.

The size and the cost assumed for each resource category therefore constitute an important part of the 'scenario variables', but first we need a concept to formulate these categories. For this purpose, we use the McKelvey diagram as introduced in Figure 2.3. That figure distinguishes eight different categories, identified by the Roman numerals I to VIII. The amounts of oil and gas in these eight categories have been taken from Rogner (1997). We summarize them in Table 5.4.

In order to further characterize oil and gas resources also in commonly used terms, Table 5.4 includes the additional distinction between conventional and non-conventional occurrences. This distinction is based on physical, technical and economic criteria. Categories I to III thus encompass conventional oil and gas quantities that can be delineated with present development practice and are amenable to the application of existing recovery technology.

In the past, on average only 34 per cent of the *in situ* oil and 70 per cent of natural gas were recovered with primary or secondary production methods. An additional fraction of the original *in situ* oil and gas can be (and already has begun to be) recovered with advanced production technologies from previously abandoned and existing fields. Category IV reflects this additional fraction, thus defining the potential for enhanced recovery of conventional resources.

Categories V to VIII encompass unconventional oil and natural gas. Unconventional oil and natural gas reserves cannot, in general, be tapped

Table 5.4 World resource base estimates by categories in ZJ (10^{21} joules)

	Conventional		Unconventional				Actual consumption, 1860–1998
	Reserves and resources	Enhanced recovery	Reserves and resources	Additional occurrences			
Category	I, II, III	IV	V, VI	VII	VIII	Total	
Oil	12.4	5.8	16.0	24.6	35.2	94.0	5.1
Gas	16.5	2.3	16.6	16.2	800.0	852.0	2.4
Coal	25.2	(n.a.)	100.3	–	–	125.5	5.5

Sources: Nakićenović *et al*. (1996, 1998), Masters *et al*. (1994), Rogner *et al.* (2000).

with conventional production methods, for technical or economic reasons, or both. Examples for unconventional oil would be oil shale, tar sands, bitumen, heavy and extra-heavy crude oils as well as deep-sea oil occurrences. Unconventional natural gas includes coal-bed methane. For a more detailed definition of these resource categories, see Rogner (1997).

Present-day *reserves* of conventional and unconventional oil and gas are defined as those occurrences that are identified, measured and, at the same time, known to be technically and economically recoverable. They are therefore the total of Categories I, II and V.

Present-day *resources* of conventional and unconventional oil and gas are defined as occurrences with less certain geographical assurance and/or with still uncertain economic feasibility. They are therefore the total of Categories III and VI.

Categories VII and VIII quantify *additional occurrences*. These are occurrences with unknown degrees of geological assurance and/or with unknown or without economic significance. An important example of natural gas in Category VIII is methane hydrates,[7] and including this category in a scenario means that technological progress will be fast enough to make the production of huge amounts of gas hydrates economically feasible before the end of the 21st century.

On the basis of this quantification, MESSAGE inputs on the total availability of resources are specified by including oil and gas categories up to a given index. For the two scenarios described in this chapter, this specification is given in Table 5.5. Total quantities of producible fossil fuels, assumptions on the technological progress of extraction and other model features concerning the dynamic availability of resources determine the shadow prices of these primary-energy carriers. Shadow prices – or marginal costs – are defined as the costs of the last unit of resource produced at a given point in time. Shadow prices are not the same as market prices, but they are an important determining factor, and their development over time provides a good description of what happens in the resource part of the MESSAGE model. We summarize the shadow prices of fossil fuel production in the two scenarios in Table 5.6.

Table 5.5 Resource categories included in MESSAGE (inputs)

Scenario	Oil	Natural gas (methane)	Coal
Post-fossil (PF)	I–VI	I–VII	I–VI
Oil and gas-rich (OG)	I–VII	I–VIII	I–VI

Table 5.6 Cost development (shadow prices) of primary fossil fuels in the PF and OG scenarios

Scenario	Cost development of primary fuels (US$/GJ)			
	2000	2020	2050	2100
Post-fossil (PF)				
Oil	3.2	4.5	5.4	4.1
Natural gas	1.9	2.1	3.1	2.0
Coal	1.0	1.0	1.3	1.0
Oil and gas-rich (OG)				
Oil	3.2	4.7	7.4	5.3
Natural gas	1.9	2.2	4.8	4.0
Coal	1.0	1.1	1.5	1.4

Shifting to higher cost categories in MESSAGE increases the price of fossil fuels, but this increase is mitigated in both scenarios by learning effects in the extraction technologies. Over the time horizon, the combined result of these two effects is, on average, a gentle increase of fossil fuel prices in the course of the 21st century. This increase is even less pronounced in the PF scenario, where the more expensive cost categories are left untapped.

With these inputs on resource categories, MESSAGE determines how much of the fossil primary energy 'resources' and 'additional occurrences' will have become 'reserves' by the end of the 21st century. With the technological development assumed for a scenario, the inclusion of a category in the model means that occurrences currently classified as resources are assumed to become reserves within the model's time horizon, whereas energy occurrences that are assumed to have little or no commercial exploration during the model's time horizon remain excluded from MESSAGE inputs.

Potentials of renewable energy sources (biomass, solar, wind and geothermal) are given in the MESSAGE model annually and for each world region. Uranium resources are given as total availabilities at the global level. MESSAGE includes the whole cycle of uranium use; that is, uranium ore extraction, its use as a fuel in nuclear power plants, and the recycling and reprocessing for re-use. Potentials of both uranium and renewable energy turn out to be ample enough to permit the widespread dissemination of these energy forms. This is the consequence not so much of perhaps assuming huge reserves but rather of limiting the speed of technological diffusion in the dynamic constraints of the model. These constraints (see also the model description in the appendix) control the capacity expansion of technologies.[8]

Technological progress of power generation and synthetic fuel technologies

Technologies included in MESSAGE are defined as time series of energy inputs, energy outputs (efficiency), costs, plant life, utilization factors and environmental impacts (emission coefficients). Technological progress of any individual energy technology is therefore represented by a favourable development of any one of these descriptors individually or in any combination with other descriptors. Since MESSAGE literally contains several hundreds of individual technologies, it would appear impractical to include here a table with time series of all technologies and all performance descriptors. We therefore chose the alternative of presenting selected descriptors of only the most important technologies.

We begin by giving, in Table 5.7, an overview of the most important aggregated electricity generation and synthetic fuel technologies represented in MESSAGE. In the table, centralized electricity generation includes those power plants that are connected to the electricity grid, whereas decentralized generation means power not connected. Synthetic fuels production technologies produce liquid or gaseous fuels from solid or gaseous fossil (coal, natural gas) or from renewable (biomass) hydrocarbons.

Although approximately 400 individual technologies are included in the MESSAGE runs describing the OG and PF scenarios, not every imaginable future technology is included in the model, and the actual choice of technologies was based on the following guiding principles. First, technologies not yet demonstrated to function on a prototype scale were excluded. Therefore, for instance, nuclear fusion was excluded. However, production of hydrogen and biomass-based synthetic fuels (for example ethanol), advanced nuclear and solar electricity generation technologies are included, as they have demonstrated their physical feasibility at least on a laboratory or prototype scale, or even in some specific niche markets.

The range of technology-specific assumptions on unit investment cost is derived from empirical distributions of technology costs taken from CO2DB, a large inventory of technology data that has been developed at IIASA (Strubegger *et al.*, 1999). Means, maximum and minimum values from these distributions (for example of estimated future technology costs) guided the decision as to which particular values to adopt under specified qualitative assumptions on technology strategies. Table 5.8 summarizes the assumptions for the electricity generation and synthetic fuel production technologies in terms of levelized costs (annualized investment and operating costs converted to specific energy output costs, excluding fuel costs).

Table 5.7 Major technologies included in MESSAGE

Technology aggregates	Description
Centralized electricity generation	
Coal, conventional	Conventional coal power plants with DESOX (e.g., fuel-gas desulphurization, FGD) and DENOX (NO_X scrubbing)
Integrated coal gasification combined-cycle (IGCC)	A technology converting coal into a gaseous fuel, which is then used as a fuel in a combined cycle; i.e. the combination of a gas turbine and a steam turbine
Coal fuel cell	Coal-based high-temperature fuel cells
Oil	New standard oil power plants (Rankine cycle, low NO_X and with FGD); existing crude oil and light oil engine-plants; light oil combined-cycle power plants
Gas, standard	Standard gas power plants (Rankine cycle, potential for cogeneration)
Natural gas-fired combined-cycle power stand (NGCC)	A combination of a gas turbine cycle and a steam turbine cycle
NGFC	Natural gas-powered high-temperature fuel cells, optional cogeneration
Biomass-fired power plant	A technology generating electricity with conventional steam turbines, through direct combustion of biomass residues, or combined-cycle gas turbines with gasified biomass (gaseous fuels thermochemically converted from biomass)
Nuclear	Conventional, existing nuclear plants
Advanced nuclear	Nuclear high-temperature (in particular 'inherently safe') reactors for electricity and hydrogen coproduction, and fast breeder reactors
Hydropower plant	Hydropower plants
Wind power plant	Wind power plants
Other renewables	Geothermal power plants (optional cogeneration); grid-connected solar photovoltaic power plants; solar thermal power plants with storage; solar thermal power plants for hydrogen production
Decentralized electricity generation	
Hydrogen fuel cell	Used in stationary or mobile applications for the combined generation of heat and power or off-peak electricity generation, generated directly from the

Table 5.7 (continued)

Technology aggregates	Description
	electrochemical reaction between hydrogen and oxygen
Photovoltaics	On-site devices that convert sunlight directly into electricity, used in the residential and commercial sectors, and in the industrial sector
Synthetic fuels	
Coal synliquids	Light-oil and methanol production from coal
Biomass synliquids	Liquid fuels (such as methanol or ethanol) produced from biomass, used mainly in the transport sector
Gas synliquids	Methanol production from natural gas
Synthetic gases	Synthetic energy gases from various sources, including biomass and coal gasification
Hydrogen, H2(l)	Hydrogen production from fossil fuels (coal or natural gas)
Hydrogen, H2(2), H2(3)	Non-fossil hydrogen production: H2(2), from biomass and electricity; H2(3), from nuclear and solar

Minimum and maximum values reflect the variability of cost assumptions over the 11 world regions.

In the OG strategy, cost reduction speeds of individual electricity generation and synthetic fuel production technologies are assumed rather modest. In contrast, cost reductions for several technologies (bio-fuel production, advanced nuclear, wind, solar photovoltaic, biomass-based synthetic-liquids and hydrogen) are assumed to proceed rather fast in the PF strategy. Cost reductions assumed for the PF strategy are on the optimistic side, compared to historical experience of the technologies' cost reductions (Grübler, 1998).

End-use technologies (useful-energy demand assumptions)
The assumptions on the technological progress of end-use technologies are given in an aggregate way, that is, as an overall energy intensity reduction, expressed in terms of declining useful energy demand per unit of GDP. The speed of decline is assumed in a way that is judged consistent with the assumed economic development; that is, we assume that higher economic growth favours steeper energy intensity reduction. Energy intensity reduction rates, together with the assumed GDP trajectory, will give the useful-energy

Table 5.8 Levelized electricity and synthetic fuel production costs (1990 US$/GJ) for selected energy technologies (excluding input fuel costs)

	1990		2050				2100			
	actual		OG		PF		OG		PF	
	min	max	min	max	min	max	min	max	min	max
Electricity generation										
Coal, conventional	3.5	7.3	4.3	7.6	4.3	7.6	4.3	7.6	4.3	7.6
IGCC	9.1	9.1	9.2	9.2	7.6	7.6	9.2	9.2	7.0	7.3
Coal fuel cell	11.6	11.6	11.3	11.3	9.5	9.5	11.3	11.3	9.2	9.2
Oil	3.8	28.1	2.2	5.1	2.2	5.1	2.2	5.1	2.2	5.1
Gas, standard	3.5	8.1	3.8	4.6	3.8	4.6	3.8	4.6	3.8	4.6
NGCC	4.7	4.9	2.2	2.7	2.2	2.7	2.2	2.7	2.2	2.7
NGFC	8.2	8.2	5.4	5.4	5.1	5.1	5.4	5.4	4.3	4.3
Biofuel	5.7	8.9	5.4	7.0	5.1	6.5	5.4	7.0	4.3	5.4
Nuclear	6.5	9.5	7.0	9.5	7.0	9.5	7.0	9.5	7.0	9.5
Adv. nuclear	10.5	10.5	9.2	12.2	6.8	10.8	8.1	10.0	4.6	8.9
Hydro	2.4	15.4	2.4	21.6	2.4	21.6	2.4	21.6	2.4	21.6
Wind	15.4	15.4	6.5	6.5	4.6	4.6	6.5	6.5	3.0	3.0
Other renewables	6.4	29.8	3.3	8.0	2.8	8.0	3.0	8.0	1.1	8.0
Hydrogen fuel cell	8.4	8.4	5.8	5.8	5.2	5.2	5.5	5.9	4.0	4.4
Photovoltaic	20.4	29.8	8.1	11.7	2.8	4.2	8.1	11.7	1.4	2.3
Synthetic fuels										
Coal synliquids	6.9	6.9	4.7	6.1	4.7	6.1	4.7	6.1	4.7	6.1
Biomass synliquids	7.1	7.1	4.0	4.0	3.1	3.1	4.0	4.0	2.4	3.1
Gas synliquids	3.7	3.7	3.6	3.6	3.7	3.7	3.6	3.6	3.7	3.7
Synthetic gases	4.6	4.6	2.9	4.7	2.9	3.1	2.9	4.7	2.8	3.1
Hydrogen H2(1)	5.6	5.6	1.4	3.3	1.4	3.2	1.4	3.3	1.0	3.2
Hydrogen H2(2)	4.9	4.9	1.5	2.9	1.3	2.8	1.5	2.9	1.0	2.8
Hydrogen H2(3)	11.9	11.9	n.a.	n.a.	3.4	10.3	n.a.	n.a.	2.9	10.3

Note: Minimum and maximum values are taken over the 11 world regions.

trajectories in seven categories: industry electricity, industry 'other' (including thermal), residential/commercial electricity, residential/commercial 'other' (including thermal), feedstock, non-commercial and transport. This calculation is done in the Scenario Generator, as described in the appendix, and the results are used as inputs to the MESSAGE model.

The numerical values of the assumed relative energy intensity are given in Table 5.9. The absolute values are graphically presented in Figure 5.4. For details on how energy intensities are related to the degree of economic development and the industrialization path that followed, see Gritsevskii (1996), who analysed cross-sectional and time-series data for many countries and estimated useful energy demand as a function of GDP per capita. The resulting quantities of useful energy demand are given in Table 5.10.

Table 5.9 Index of energy intensity (useful energy demand divided by GDP) development in two scenarios (1990 = 100)

	1990	2020	2040	2060	2080	2100
Oil and gas-rich (OG)	100	76.6	62.0	54.9	47.5	37.8
Post-fossil (PF)	100	78.8	58.4	47.4	40.3	32.8

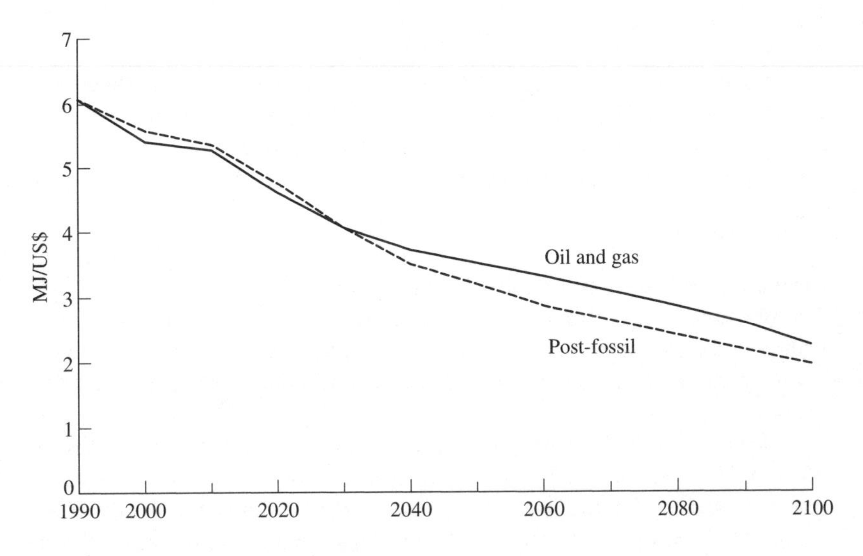

Figure 5.4 Energy intensity development, 1990–2100, global values in two scenarios (MJ per 1990 US$)

Table 5.10 Useful energy demand in two scenarios, exajoules

	1990 (Actual)	2020	2040	2060	2080	2100
Oil and gas-rich (OG)	127	265	509	847	1134	1261
Post-fossil (PF)	127	272	479	729	959	1090

5.2 MODEL RESULTS

In this section we present the MESSAGE results for the 'oil and gas' (OG) and 'post-fossil' (PF) scenarios. For the presentation of these results, we look at the two strategies from the following three perspectives. First, we look at the development of the energy system in terms of the technology mix of energy supply. In particular, we will focus on the energy technology structure observed at three stages of the energy system, primary-resource consumption, electricity generation and final-energy demand. Second, we look at the environmental implications of the technology strategies, focusing on CO_2 emissions and sulphur emissions. Finally, cost implications of the two strategies will be shown and discussed.

5.2.1 Energy Systems

In this subsection, the energy systems in the two scenarios – as they result from the alternative assumptions regarding different technology strategies – are presented in terms of energy forms used at different stages of the energy system, namely primary-energy mix, electricity and synthetic fuel output mix and final-demand mix.

Primary-energy mix, cumulative resource consumption

The primary-energy mix of a given scenario describes how much of each primary-energy form is extracted from nature. We look at the following primary-energy forms: coal, gas, oil, nuclear, biomass and other renewable energy.

As far as fossil fuels are concerned, the primary-energy mix bears directly on cumulative resource consumption. As mentioned above, resource recoverability was one of the input assumptions that differed for the two scenarios. This input limits the available resource that can eventually be converted into reserves and recovered during the model's time horizon. The actual resources consumption is the calculated result of the model.

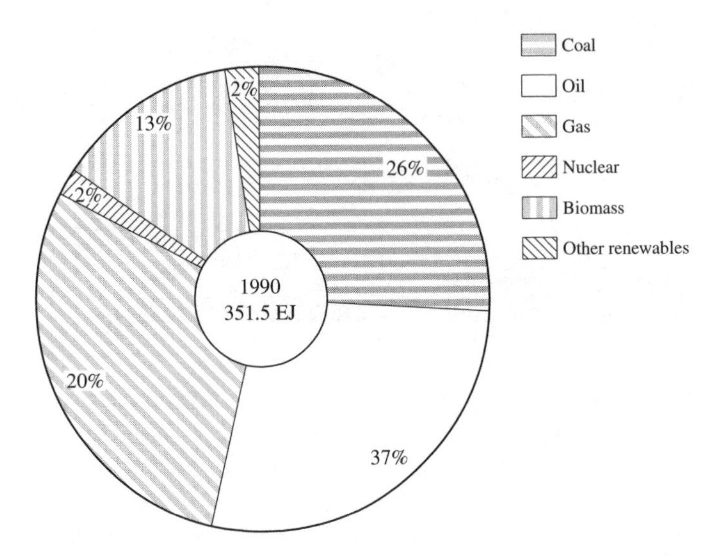

Figure 5.5 Percentage shares of primary-energy supply (total in EJ), in 1990

Between 1900 and 1990, total global primary-energy consumption, on average, increased at slightly more than 2 per cent per year while the consumption of fossil energy rose at an average annual rate of almost 3 per cent during the same time period. In 1990, 350 EJ of primary energy[9] were used globally, of which 76 per cent was fossil fuels. Biomass accounted for 13 per cent and all renewable energy together for 15 per cent (Figure 5.5). Note that biomass use in 1990 was predominantly non-commercial use in developing regions.

The projected development of the global primary-energy mix between 1990 and 2100 is illustrated in Figure 5.6 for both scenarios. In OG, global primary-energy consumption in 2050 increase by a factor of 4.2 from 1990, and by a factor of 7.8 (to more than 2700 EJ) in 2100. This corresponds to an average annual growth rate of just below 2 per cent during the 21st century, which is in line with past trends. The most striking feature of this development is the expansion of the share of natural gas, which increases from 20 per cent in 1990 to 45 per cent in 2100 (see also Table 5.11). During the same time period, the shares of nuclear and 'other renewables' are projected to increase from 2 per cent to 12 per cent and from 2 per cent to 10 per cent, respectively. These increases are accompanied by decreases in coal (25 per cent to 3 per cent) and oil (37 per cent to 14 per cent).

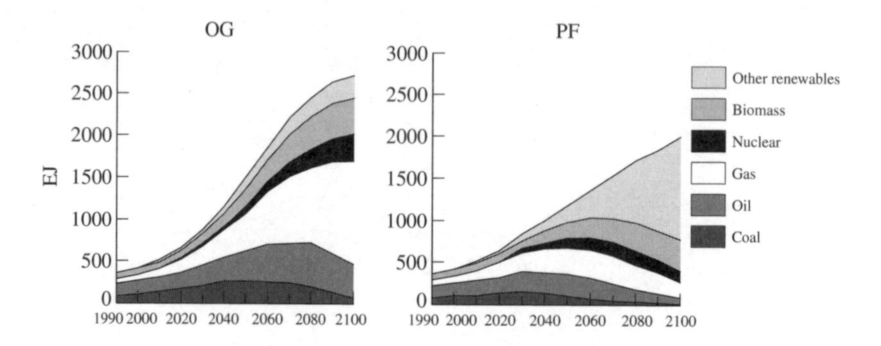

Figure 5.6 Primary-energy consumption, 1990–2100 in EJ, OG and PF strategies

Table 5.11 Percentage shares of primary-energy sources in global primary-energy supply

	Coal	Oil	Gas	Nuclear	Biomass	Other renewables
Actual (1990)	25.9	36.5	20.1	2.1	13.1	2.4
OG (2100)	3.1	14.3	45.3	12.1	15.1	10.1
PF (2100)	1.2	3.8	9.7	5.6	18.3	61.3

In contrast to OG, the increase in total global primary-energy use in the PF scenario is more modest. This is a consequence of the enhanced energy conservation and higher efficiency improvements assumed there. Global primary-energy demand in PF increases to no more than 2000 EJ in 2100, which means an increase by a factor of 5.8 from 1990 and to a level that is about one-quarter lower than that of the OG scenario. This development is equivalent to an annual average growth rate of 1.6 per cent between 1990 and 2100, which is significantly lower than the historical trend.

As to the primary-energy mix, the major feature of the PF strategy is a steep increase in the 'other renewables', to 61 per cent of total primary energy in the year 2100. This increase comes primarily from high shares of solar energy for hydrogen production and at the expense of decreasing shares of fossil fuels, in particular coal (to 1.2 per cent in 2100) and oil (to 3.8 per cent). 'Other renewables' thus become the clearly dominant primary energy source in the PF strategy in 2100, even surpassing the share of natural gas in the OG strategy (45 per cent) in the same year.

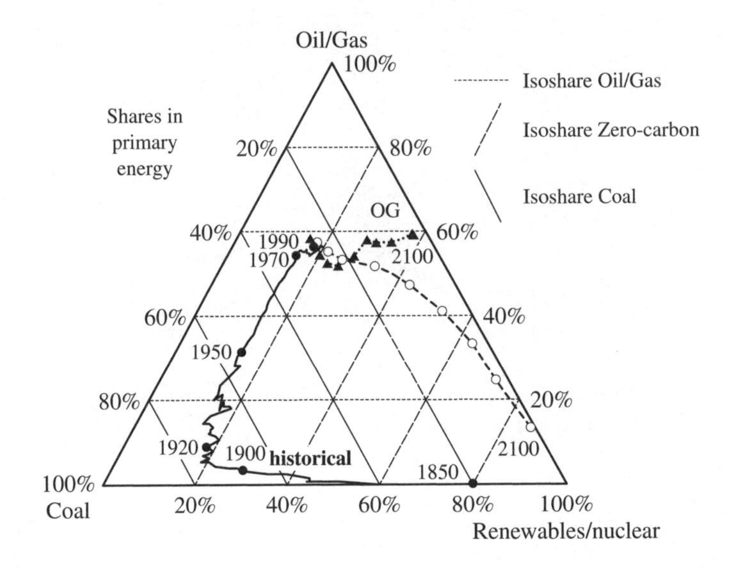

Figure 5.7 Primary-energy mix: actual development 1850–1990, scenarios 1990–2100

Figure 5.7 illustrates how the primary-energy shares have developed since 1850 and how they are projected to evolve in the two cases between 1990 and 2100. It can be seen that, in OG, the share of oil and gas remains almost the same from 1990 onwards, while the share of coal is gradually replaced by renewable energy and nuclear. In PF, existing shares of fossil fuels are mostly replaced by renewable and nuclear energy, which results in a very high share of carbon-free energy.

Figure 5.8 highlights the difference between the primary-energy mixes in 2100 for the two scenarios. The big share of natural gas in OG makes the primary-energy structure lopsided towards the vertex representing fossil fuels, whereas PF is dominated by renewable energy. The environmental implications of this difference will be discussed below in a quantitative way. We just note here that it is obvious that OG has much larger CO_2 emissions than PF.

To further characterize the primary-energy mix from renewable sources we show a disaggregation of 'other renewables' into solar energy, waste, biomass, geothermal (*Geoth.*), wind, hydropower and on-site technologies[10] (*On-site*) in Figure 5.9. Solar energy includes hydrogen that produces electricity production with fuel cells (*solar_h2*) and solar power that generates electricity directly. As we have explained above (see note 9), the

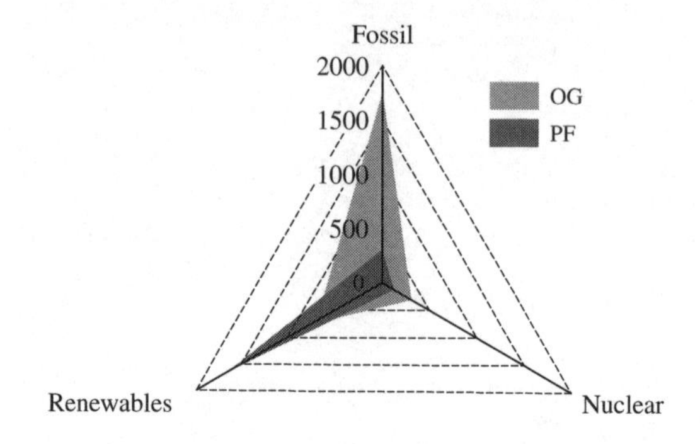

Figure 5.8 Primary-energy consumption in two scenarios in 2100 (in EJ) by three categories

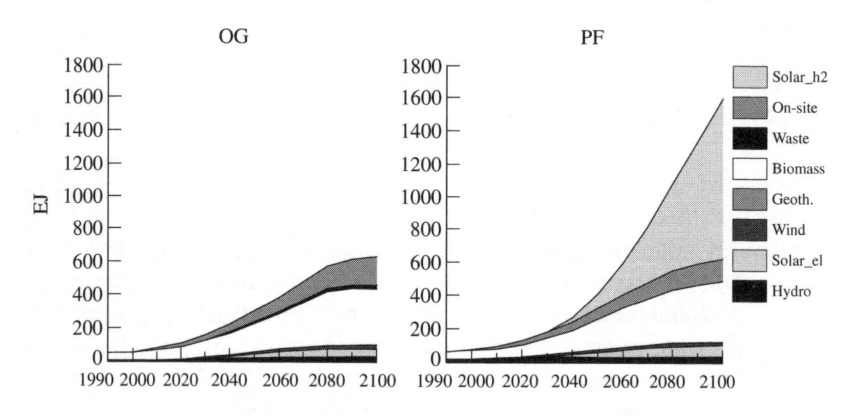

Figure 5.9 Renewable primary-energy consumption, 1990–2100 (in EJ), OG and PF

primary energy in the case of renewable energy is accounted for by using the final-energy output of these technologies.

The major difference between the two strategies with respect to primary-energy consumption by renewable energy is mainly accounted for by hydrogen produced from solar energy (*solar_h2*), which amounts to almost 1000 EJ in 2100 in PF.

Table 5.12 shows the consumption of oil and natural gas in 2100 and their respective reserves-to-production (R/P) ratios. The R/P ratio is an indicator measuring reserves in terms of current annual production under the hypothetical assumption that future production remains constant. The

Table 5.12 Oil and gas consumption (in EJ) and reserves-to-production ratios, 2100

	Natural gas		Oil	
	Consumption (EJ/yr)	R/P ratio (years)	Consumption (EJ/yr)	R/P ratio (years)
OG	1241	629	391	59
PF	196	127	77	178
1990	*72*	*58*	*139*	*43*

Note: For comparison, the last row of the table shows the corresponding 1990 values.

R/P ratio is therefore measured in years. Remaining reserves in the year 2100 are defined as the unused amounts of oil and gas in their respective MESSAGE categories (see the definition of resource categories included in MESSAGE in Table 5.5).

In both scenarios, and for both oil and natural gas, the R/P ratios in 2100 are significantly higher than in 1990. Maybe the most interesting number in the table concerns the R/P ratio for natural gas in the OG strategy, which is in excess of 600 years even though natural gas consumption in that strategy is more than 1000 EJ in 2100.[11] This is the result of assuming that, by the end of the 21st century, it will have become economically feasible to extract methane captured in methane hydrates (see pages 117–20).

By definition, reserves-to-production ratios reflect two quantities at the same time. High values of this indicator can therefore mean either large reserves or small production. The latter is the explanation for the high R/P ratio for oil in the PF strategy. There oil production in 2100 is approximately just half of what it was in the year 2000.

The cumulative use of fossil resources between 1990 and 2100 is illustrated in Figure 5.10. The figure shows that the two scenarios are similar in terms of the shares of the three fossil energy sources in the total consumption of fossil primary energy. The absolute numbers behind these shares reveal that the total production of each of these energy carriers in OG is bigger than that of PF by factors of between 1.6 and 1.9.

The same kind of information, disaggregated into the four world regions, is presented in Figure 5.11. The difference in the global resource use between two scenarios comes mainly from Asia and the 'Rest of the world' (ROW). In these regions, the major difference is oil consumption. In the OECD region, although the difference of the two strategies in absolute terms is smaller than in Asia and ROW, the PF strategy results in a 40 per cent

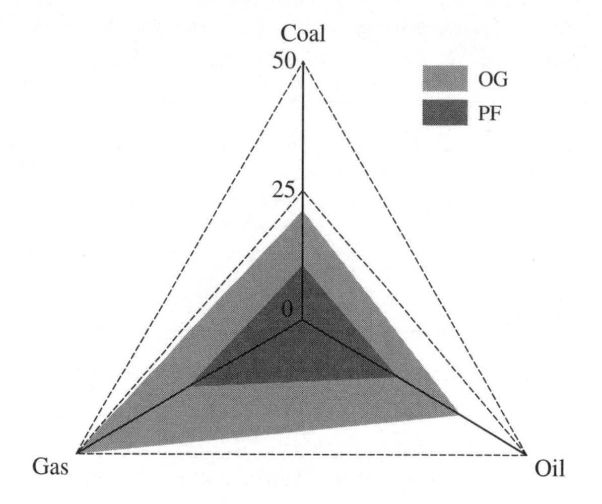

Figure 5.10 *Cumulative resource use, 1990–2100, OG and PF strategies, ZJ (10^{21} Joule)*

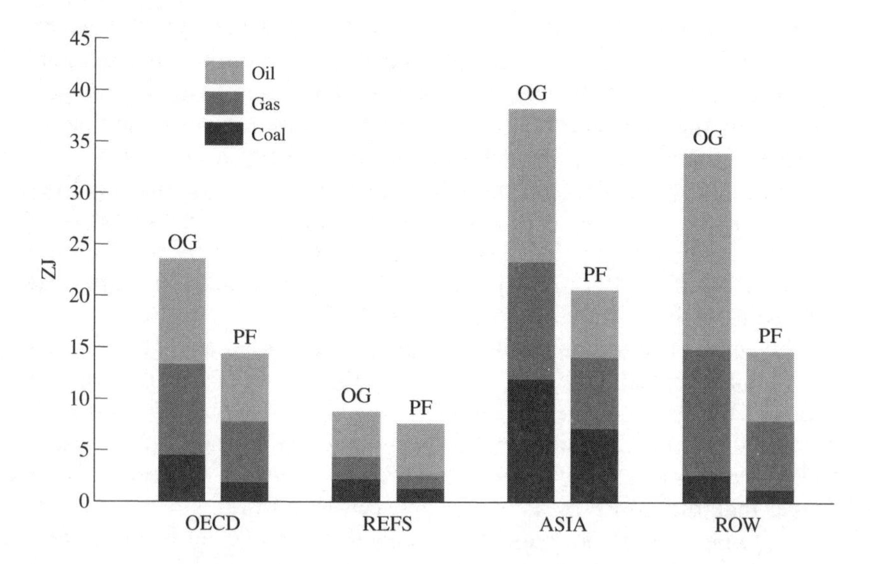

Figure 5.11 *Cumulative resource use, 1990–2100, OG and PF, ZJ (10^{21} Joule)*

smaller total cumulative use of fossil resources than the OG strategy. In Reforming Economies (REFS), the difference between the two strategies is rather small.

Electricity generation and synthetic fuels

To be suitable for consumption, primary energy has to be converted into more readily usable forms, such as electricity, synthetic liquid fuels and gases. These fuels belong to the secondary energy stage. Electricity is generated in many different ways and using many different primary energy sources. Currently, most of the global electricity is produced in a centralized way, but it can also be generated locally, such as by solar panels, gas mini-turbines and mobile fuel cells (in cars) for example. These decentralized electricity generation technologies are regarded as particularly promising technology options for future electricity generation.

Synthetic fuels play an important role in future energy systems, primarily for two reasons. First, they may replace oil as the main transportation fuel and, secondly, they pollute the environment less than oil products. This is particularly true of hydrogen, which produces almost no pollution at the point of its final use.

Electricity generation Figure 5.12 shows the shares of several primary-energy carriers in global electricity generation in 1990. In that year, 63 per cent of global electricity was generated from fossil fuels, of which coal accounted for the major part (38 per cent of the total). 'Other renewables', consisting mainly of hydropower, and nuclear energy supplied most of the balance. The share of biomass was small (1 per cent) and the share of decentralized electricity generation was zero.

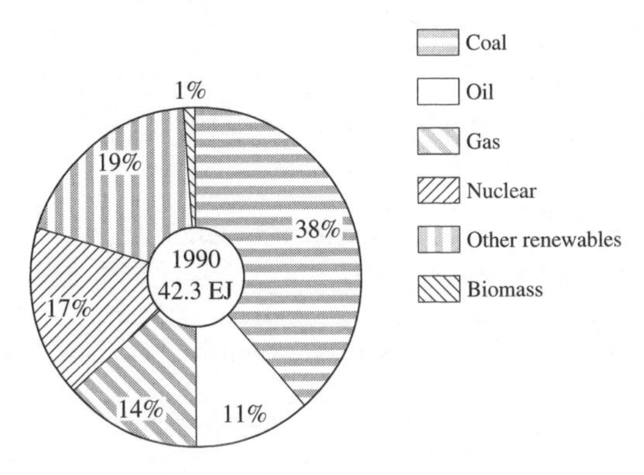

Figure 5.12 Percentage shares of global electricity generation (total in EJ), 1990

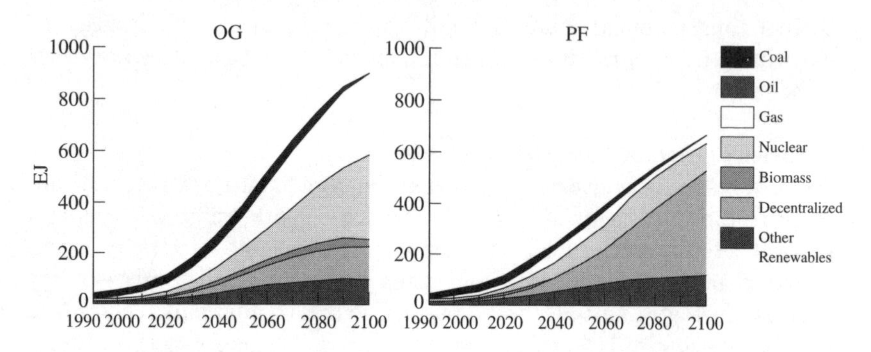

Figure 5.13 Electricity generation mix, OG and PF strategies (EJ_{el}/yr)

Figure 5.13 describes the evolvement of the global electricity generation mix between 1990 and 2100. The figure shows that total electricity generation in 2100 is more than 20 per cent lower in PF than in OG. This is a consequence of the assumption that technological progress in PF will lead to a steeper reduction of final-energy intensities in this scenario than in OG.

Particularly in the second half of the 21st century, the difference between the two scenarios is characterized by the use of different primary-energy forms and by different degrees of centralization of the power generation system. The shares of each electricity generation technology in the total generation in the year 2100 are summarized in Table 5.13.

In OG, natural gas and nuclear energy become the two biggest sources of electricity generation. The contribution from natural gas to global electricity generation expands gradually from its current level of 14 per cent to 35 per cent in 2100. Nuclear energy expands steadily to reach 36 per cent by the end of the century, thus more than doubling its 1990 share of 16.5 per cent. By that time also, hardly any electricity is generated by coal or oil.

The PF strategy is characterized by a big share (almost 80 per cent) of

Table 5.13 Shares of global electricity generation in 2100 and actual 1990 values(per cent)

	Coal	Oil	Gas	Nuclear	Biomass	Decentralized	Other renewables
Actual (1990)	38.5	11.4	13.6	16.5	1.1	0	18.9
OG (2100)	0.2	0	34.7	36.2	3.0	2.3	11.6
PF (2100)	0.0	0	4.4	15.8	0.9	62.0	16.9

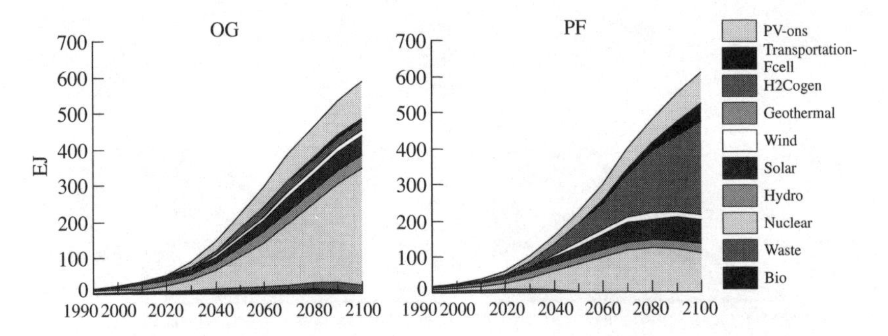

Figure 5.14 Electricity generation from non-fossil primary-energy sources in the OG and PF strategies (in exajoules, EJ_{el}/yr)

renewable energy (sum of biomass, decentralized and 'other renewables') in electricity production, in particular decentralized generation (62 per cent). Electricity generation by coal and oil disappears completely, and the share of natural gas decreases to 4.4 per cent of the total. In both scenarios, carbon-free energy carriers dominate global electricity generation in the long run. We therefore show, in Figure 5.14, the electricity generation by nuclear and renewable energy carriers only. The first point to note here is that total generation by carbon-free sources is higher in PF than in OG, despite the fact that OG has the higher generation by all sources taken together. A noteworthy feature of the generation mix is that the share of nuclear energy is much higher in OG than in PF. Moreover, in OG, nuclear electricity generation increases at the end of the model's time horizon, whereas in PF it decreases. The increasing and dominant source of electricity in PF is the cogeneration of heat and electricity using hydrogen in stationary fuel cells. This marks the PF strategy as one with a highly decentralized energy system.

Synthetic fuels The primary importance of synthetic liquid fuels in our scenarios is that they replace oil in the transport sector. Therefore we focus the presentation here on the final-energy supply mix in this sector (Figure 5.15). Methanol and ethanol are important final-energy carriers in both scenarios. In the oil and gas strategy (OG), the supply by these two synthetic fuels reaches some 40 per cent of total sectoral supply by 2100. In PF, hydrogen becomes important in addition, supplying more than 100 EJ of transport energy in 2100. In PF, synthetic fuels thus supply 75 per cent of the transport demand by the end of the 21st century.

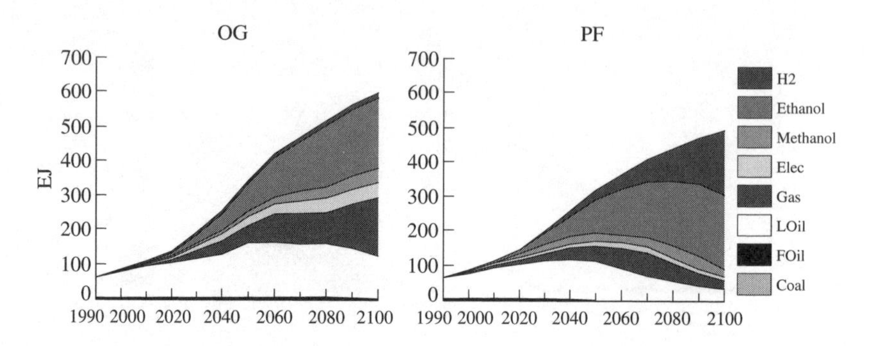

Figure 5.15 Final-energy supply in the transport sector in two scenarios (EJ/yr)

Final-energy supply mix

Final-energy is the energy form that could be considered the most important from an economist's point of view because it is the main energy form that fulfils the economic function of producing output for final consumption. In MESSAGE, final-energy demands are calculated from the optimized supply of (given) useful-energy demand.

Figure 5.16 shows global final-energy consumption by energy form in 1990, disaggregated by final-energy fuel categories. Total final-energy consumption in 1990 was 275 EJ, with oil products accounting for as much as 40 per cent of it. All fossil final energy together contributed 68 per cent of the total, whereas electricity took no more than a 13 per cent share.

In OG, global final-energy consumption increases by a factor of 3.7 between 1990 and 2050, corresponding to an annual average growth rate (AAGR) of 2.2 per cent, and by as much as a factor of 6.4 (an AAGR of 1.7 per cent), to 1766 EJ, in 2100. The evolvement of the final-energy supply mix in the two scenarios is shown in Figure 5.17. The shares of the final energy use in 1990, and in the OG and PF scenarios for 2100 are presented in Figure 5.18.

The figure illustrates that, by 2100, several new energy forms contribute to final-energy supply in both scenarios. The OG scenario is still characterized by a high share of electricity (nearly 40 per cent). Fossil-based final energy (coal, oil products and natural gas) is reduced to a 30 per cent share, from 80 per cent in 1990. On-site electricity generation from photovoltaic conversion and synthetic liquids (predominantly hydrogen but, to some extent, also alcohols) grow steadily from zero in 1990 to 11.1 per cent and 11.7 per cent, respectively, in 2100.

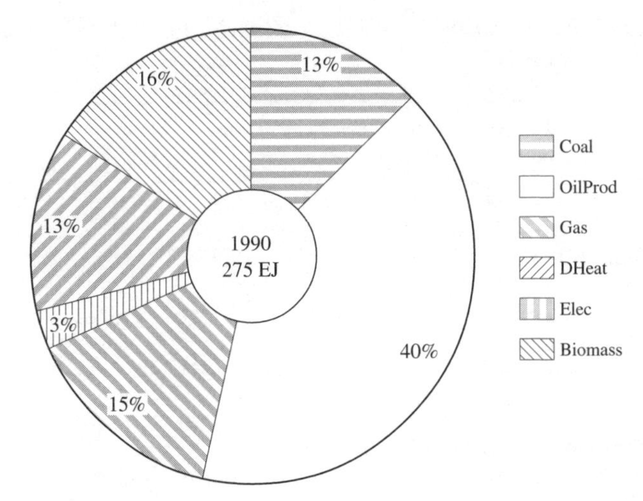

Figure 5.16 *Global final-energy consumption by energy form in 1990, percentage shares, total in exajoules*

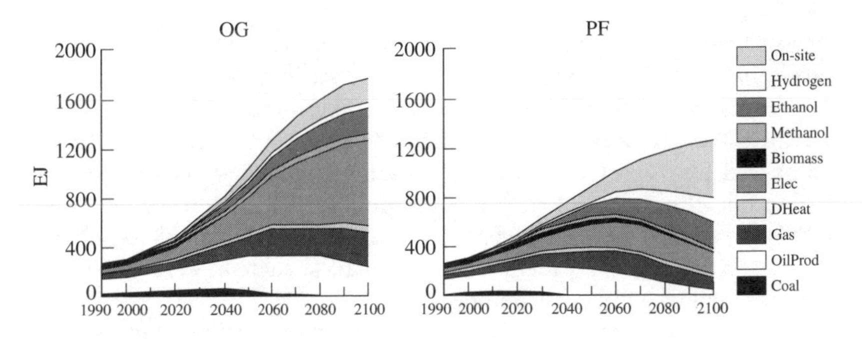

Figure 5.17 *Final-energy mix, OG and PF scenarios (in EJ/yr)*

In the PF scenario, total global final-energy use increases to not more than 1270 EJ in 2100, which is a factor of 4.6 (equivalent to an AAGR of 1.4 per cent) from 1990 and 28 per cent lower than OG's value in the same year. For the first half of the 21st century, the increase in final-energy demand is accounted for by increases in electricity, ethanol and on-site electricity generation. After 2050, demand for grid-delivered electricity ceases to grow, whereas in particular on-site electricity generation by photovoltaic conversion and synthetic liquids (predominantly ethanol and hydrogen) continue to grow rapidly until the end of the century, and gain as much as 35 per cent and 37 per cent of the market share, respectively. The shares of final-energy supply in the year 2100 are summarized in Table 5.14.

Table 5.14 Final-energy shares in two scenarios in 2100 and actual shares in 1990 (%)

	Coal	Oil prod.	Gas	District Heat	Electricity	Biomass	Methanol ethanol	Hydrogen	On-site
Actual (1990)	13.1	40.4	14.9	2.8	12.7	16.2	0	0	0
OG (2100)	0.1	13.5	16.3	3.0	38.7	0.1	15.2	2.0	11.1
PF (2100)	0.0	4.6	7.9	1.8	13.9	0.1	20.0	15.2	36.5

World-regional energy systems in 2100

Let us now look into all 11 world regions of our study. For the sake of brevity, our presentation will focus on snapshot pictures of the 11 world-regional energy systems for the years 1990 and 2100, rather than on the development over the entire time span of a hundred years.

Primary-energy mix, cumulative resource consumption The primary-energy mix in 1990 is shown in Figure 5.18 for the 11 world regions. In that year, 24 per cent of the global primary energy was consumed in North America (NAM). Western Europe (WEU) and the Former Soviet Union (FSU) accounted for 16 per cent each, and Centrally Planned Asia and China (CPA) for 10 per cent of global consumption. None of the other world regions consumed more than 6 per cent of the global total, and Sub-Saharan Africa (AFR), the world region with the smallest share in 1990, accounted for 3 per cent of global consumption.

As to the patterns of primary-energy mixes in the world regions in 1990, NAM, WEU and Pacific OECD (PAO) had similar consumption patterns, in which oil consumption accounted for the largest part, with coal and gas dominating the remainder. Eastern Europe (EEU) depended on coal to a larger extent, whereas FSU depended mainly on natural gas. In Latin America (LAM), Pacific Asia (PAS), and Middle East and Northern Africa (MEA), coal was not an important source. Biomass was important in LAM and PAS (accounting for 30 per cent and 37 per cent of consumption, respectively). Not surprisingly, dependence on oil was rather high (65 per cent) in oil-rich MEA. In South Asia (SAS) and AFR, biomass accounted for half of the regions' primary-energy consumption. The consumption of natural gas was insignificant in both of these regions.

Figure 5.19 shows the world regional primary-energy mixes for the year 2100 for the two scenarios. The left column for each region corresponds to

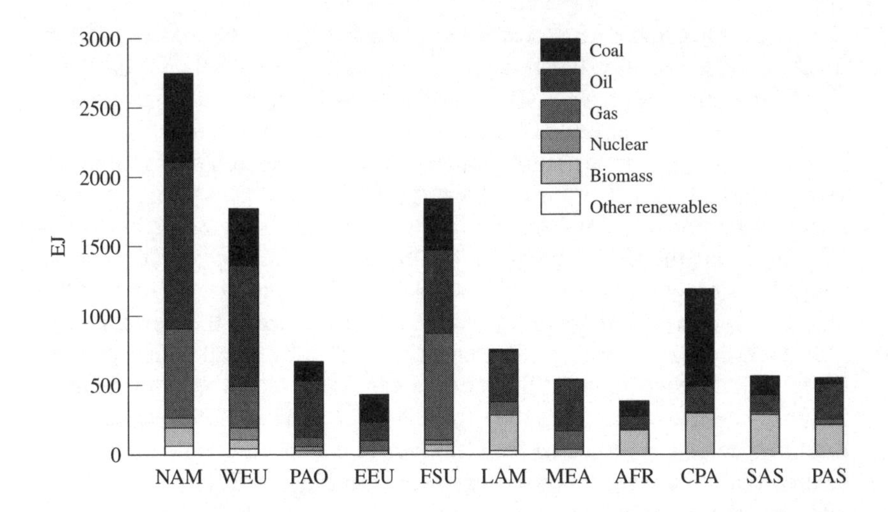

Figure 5.18 Primary-energy consumption (EJ) in 11 world regions, 1990

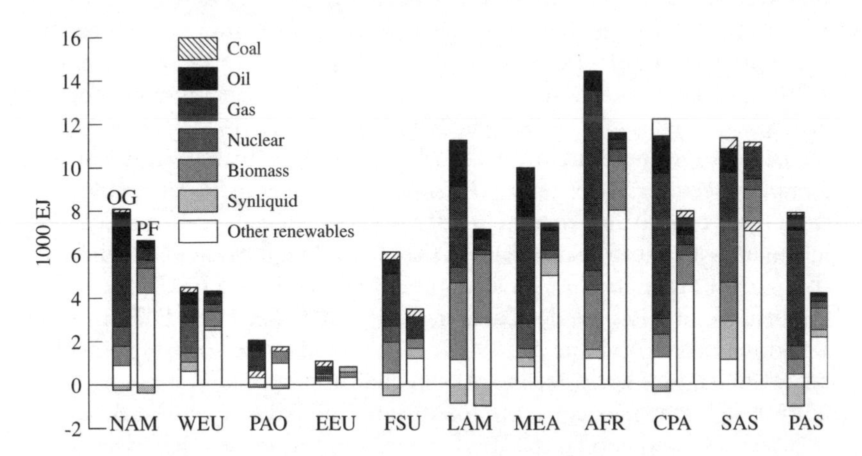

Note: For each world region, the left column shows the result for the OG strategy, the right column for the PF strategy.

Figure 5.19 Primary-energy consumption (EJ) in 11 world regions, 2100

OG, and the right one to PF. The world-regional shares of primary-energy consumption are quite similar for the two scenarios. Three world regions (NAM, WEU and FSU) that together accounted for 56 per cent of the global primary-energy consumption in 1990 account for only 18 per cent (OG) and 15 per cent (PF) respectively in 2100. In contrast, the added shares for AFR and SAS increase significantly (for OG to 30 per cent and for PF to 40 per cent of the global total compared with 8 per cent in 1990). In OG, the shares for LAM, MEA, CPA and PAS also increase significantly (to 12 per cent, 12 per cent, 14 per cent and 8 per cent respectively).

The overall global pattern of primary-energy mix is well mirrored in the world-regional shares of primary-energy mix in 2100: gas consumption increases significantly for OG, and renewable energy is increasingly used in PF. Note that the negative entries for 'Synliquid' are the primary-energy equivalents of synthetic fuel exports. They are added to give an idea of the amounts of energy that are produced but not consumed in the world region.

In OG, coal consumption decreases substantially in all world regions except EEU where coal still accounts for 21 per cent of the region's primary-energy supply in 2100 (from 48 per cent in 1990). The most dramatic decline is projected for CPA, where coal accounts for only 7 per cent of that region's primary-energy supply in the year 2100 – down from a 60 per cent share in 1990.

Oil continues to be an important source of primary energy in many world regions, but to a significantly lesser degree than in 1990. Instead of oil, which in 1990 was the biggest single primary-energy source, gas becomes the biggest source of primary-energy supply in the OG strategy. Except for WEU and FSU, the share of gas in total primary-energy consumption grows, particularly in developing countries. In PAS, the share of gas increases from 7 per cent in 1990 to 78 per cent in 2100; in AFR it increases from 1 per cent to 57 per cent; in CPA from 1 per cent to 36 per cent; in SAS from 6 per cent to 20 per cent; in PAO from 12 per cent to 49 per cent.

Biomass use increases its share in those world regions that had comparatively small shares in 1990, for example in PAO, EEU and FSU. The relative importance of biomass decreases in AFR, CPA, SAS and PAS. In some world regions, NAM and WEU in particular, increased shares of nuclear energy make up for the decline of coal and oil supply.

In the PF scenario, consumption of both oil and coal almost disappear in all world regions. The shares of gas consumption also decrease significantly, although in FSU and SAS natural gas still maintains some importance in the primary-energy supply.

In developed regions, the importance of renewable energy increases significantly, from negligible levels in 1990 to more than 60 per cent in 2100. In particular, hydrogen, generated from carbon-free sources (H2_0C) becomes a very important final energy source. In NAM, WEU, PAO, MEA, AFR and SAS, the shares of primary energy used to produce hydrogen in this way become higher than 40 per cent (but below 50 per cent) of final-energy supply. In developed countries also biomass increases its importance as a primary-energy source.

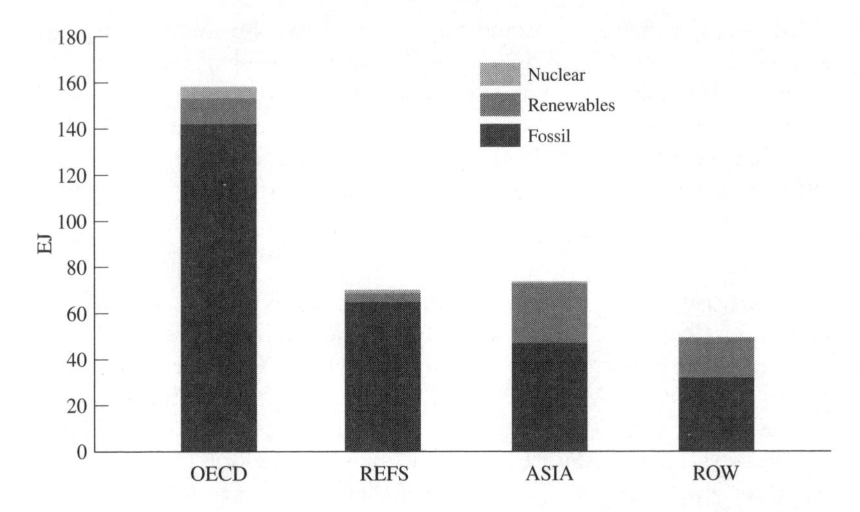

Figure 5.20 Primary-energy use (EJ) by three categories and four world regions, 1990

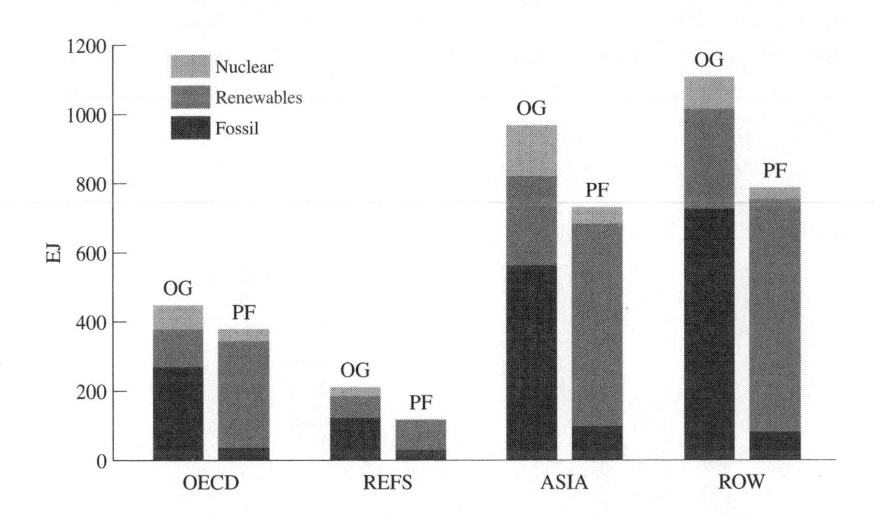

Figure 5.21 Primary-energy use (EJ) by three categories and four world regions, 2100

Figures 5.20 and 5.21 summarize the primary-energy mix changes in the scenarios as they have just been described, now aggregated for the four aggregated world regions, OECD, Reforming Economies (REFS), Asia and Rest of the World as well as aggregated by kind of fuel. The two figures clearly show shifts from a dominance of fossil-based primary-energy consumption in 1990 (Figure 5.20) to a less fossil-dependent pattern in 2100 (Figure 5.21). This feature is particularly pronounced in PF. In all world regions, the PF scenario shows a radical shift into primary-energy supply by renewable energy carriers. In OG, the most notable feature is the increased share of nuclear energy in non-OECD regions, particularly in the Asian region.

Electricity generation Figure 5.22 shows the supply mix of electricity generation in 1990 in the 11 world regions of our study. Similar to the geographical distribution of the primary-energy use, the developed regions North America (NAM) and Western Europe (WEU) generated 60 per cent of the global electricity. The two regions in transition, Former Soviet Union (FSU) and Eastern Europe (EEU), accounted for 19 per cent, and the other seven, developing, regions together for 20 per cent of global electricity generation.

In 1990, coal-based electricity generation had the largest share of electricity production in Sub-Saharan Africa (AFR) with 72 per cent, followed by China (CPA) with 71 per cent, EEU with 64 per cent, South Asia (SAS) with 56 per cent and NAM with 49 per cent. In WEU, coal-based electricity generation accounts for the greatest share of electricity generation (34 per cent), but nuclear-based electricity generation also has a large share

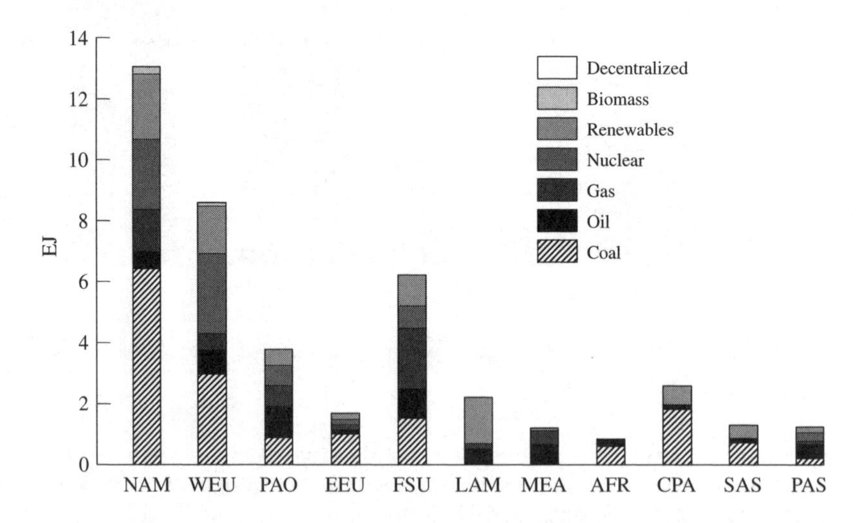

Figure 5.22 Electricity generation mix (EJ_{el}), 1990

(31 per cent), which amounts to 25 per cent of the world's total nuclear power generation.

In Pacific OECD (PAO), half of the total electricity is generated by oil and coal. Nuclear also plays an important role in that region, accounting for 20 per cent of the electricity generation. Latin America (LAM) has a unique electricity generation mix in comparison to other regions. Here 65 per cent of the electricity is generated by renewable electricity generation, accounting for 30 per cent of the global renewable-based electricity generation. No other world regions have such a high share of electricity generated by renewable sources. Other regions that have a relatively high share of renewable electricity generations are SAS (28 per cent), CPA (23 per cent) and WEU (19 per cent). Electricity generated by biomass has negligible shares in all regions.

Figure 5.23 shows the world-regional distribution of the electricity generation in 2100 for the two scenarios. The biggest differences between the scenarios are due to the higher hydrogen production in PF, which reduces electricity demand in comparison to OG. In OG, which is described by the left bar of the two, AFR and SAS generate the largest amounts (16 per cent and 15 per cent, respectively) of global electricity. In PF, CPA will become by far the biggest generator of electricity, accounting for 28 per cent of the global electricity generation.

In the OG scenario, the electricity generation mix in all world regions is characterized by an increasing dependence on natural gas and nuclear energy. PAO, AFR and CPA produce more than half of their electricity with natural gas (66 per cent, 54 per cent and 51 per cent, respectively).

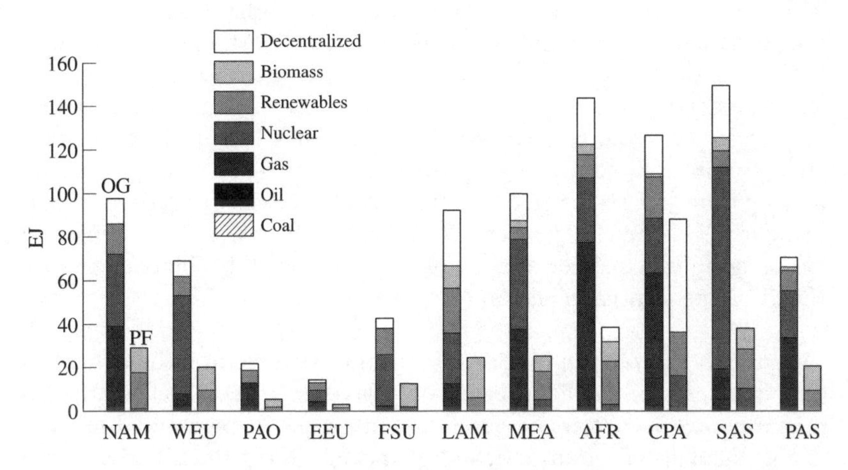

Figure 5.23 Electricity generation mix, EJ of secondary energy, 2100

Coincidentally, these are the world regions that had negligible natural gas shares in 1990, and thus the change in the electricity generation mix in the year 2100 is quite significant. NAM and EEU will also see a considerable increase in the share of gas-based electricity generation (both to 42 per cent). An exception is FSU, where gas-based electricity generation decreases from 33 per cent in 1990 to 8 per cent in 2100.[12]

Regarding nuclear energy, WEU, FSU and SAS have high shares of nuclear-based electricity generation in 2100, accounting for more than half of the total electricity generation in each of these three world regions (66 per cent, 54 per cent and 51 per cent, respectively). In all world regions except PAO, the reliance on nuclear energy increases to shares from 20 to more than 30 per cent. The reason for the exception is that PAO has relatively high reserves of unconventional natural gas, which are used for electricity generation and hydrogen production in preference to nuclear.

Another notable feature of the electricity generation mix in 2100 in the OG strategy is the increased importance of decentralized electricity generation in all regions. In most world regions, it accounts for approximately 15 per cent of the region's total electricity generation, with the exception of PAS (6 per cent) and LAM (27 per cent). In 2100, LAM has the biggest share of global decentralized electricity generation.

In the PF scenario, biomass and 'other renewable' energy become very important. In most world regions, these two primary-energy sources together account for 70 per cent of total electricity generation. WEU, PAS, EEU, LAM and NAM are projected to produce more than 95 per cent of electricity generation by renewable sources including biomass (100 per cent, 100 per cent, 98 per cent, 95 per cent and 95 per cent, respectively). In NAM, MEA, AFR and SAS, the relative importance of 'other renewable' energy is higher than that of biomass, whereas in PAO, LAM and FSU, biomass has higher shares. Note that, in FSU, electricity generation by renewable energy does not play a role. Instead, nuclear energy accounts for the balance of electricity generation (25 per cent). In MEA and SAS, nuclear also plays an important role, accounting for 16 per cent and 31 per cent of the electricity generation, respectively. A unique mix of the electricity generation is found in CPA. There, 60 per cent of the total electricity in 2100 is generated in a decentralized manner. The rest of the generation is by 'other renewable' energy (23 per cent) and nuclear (17 per cent).

The transport sector In MESSAGE, the transport sector includes light oil (gasoline, kerosene and light heating oil, alias diesel), heavy fuel oil, which is primarily used for electricity generation in big power plants, and to some minor extent also for transport, and synthetic fuels ('synfuels'), which comprise alcohols and hydrogen.

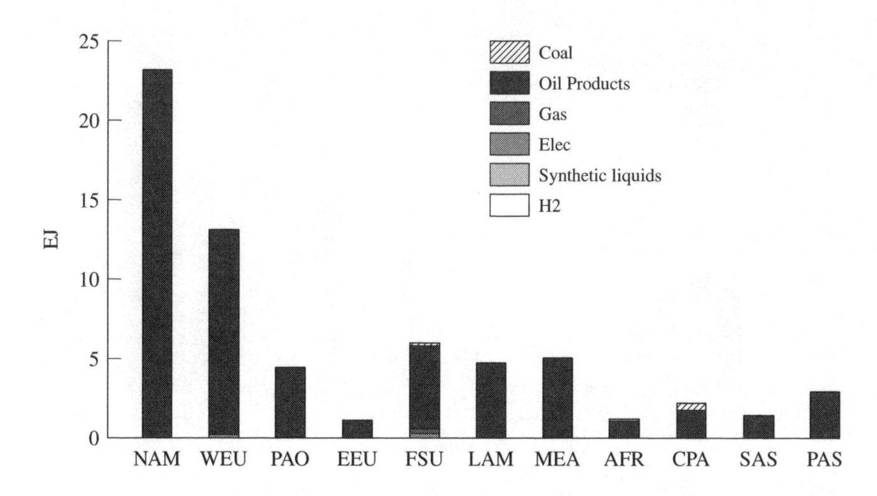

Figure 5.24 Final energy supply (EJ) in the transport sector, 1990

Figure 5.24 shows final-energy supply in the transport sector in 1990. Then, supply was concentrated in the industrialized world regions with North America (NAM), Western Europe (WEU) and Pacific OECD (PAO) accounting for more than 60 per cent of the global supply in this sector. The final-energy supply mix was homogeneous: oil products (light oil plus fuel oil) supplied more than 90 per cent of transport demands in most world regions. A minor exception is Centrally Planned Asia and China (CPA), where 20 per cent of demand was supplied by coal.

Figure 5.25 shows world-regional final-energy supply in the transport sector in 2100 in the two scenarios. The figure shows major demand increases in developing world regions in both strategies, in particular in Sub-Saharan Africa (AFR), South Asia (SAS), Centrally Planned Asia and China (CPA), as well as in Middle East and Northern Africa (MEA). In the OECD world regions (North America, NAM), Western Europe, WEU, and Pacific OECD, (PAO) and the transition countries (Eastern Europe, EEU and Former Soviet Union, FSU), total transport demands do not show major differences between the two strategies, OG and PF. The global difference in transport energy demand between the two strategies comes from the developing regions, particularly from AFR, CPA, Latin America (LAM) and Pacific Asia (PAS).

A major characterization of the final-energy supply mix in the transport sector of both scenarios in 2100 is the utilization of new forms of energy and a shift away from the domination of oil. In the PF strategy, hydrogen (H2) becomes an important transport fuel in NAM, MEA, AFR, PAO and

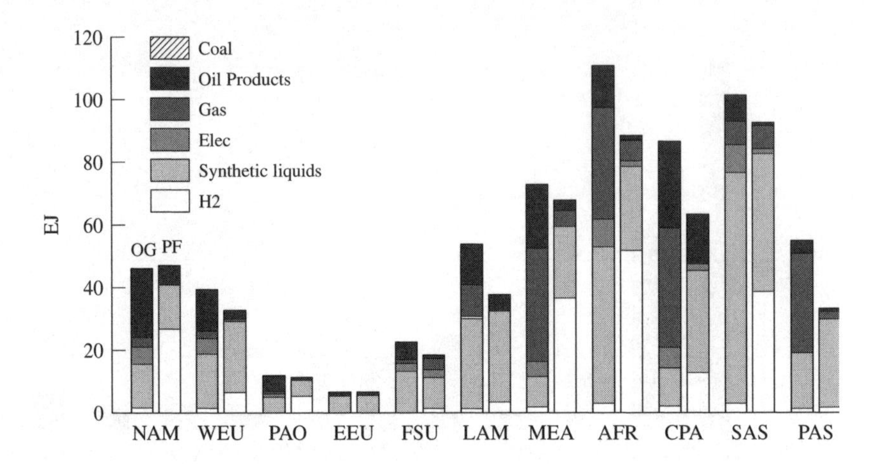

Figure 5.25 Final energy supply (EJ) in the transport sector in two scenarios, 2100

SAS, where, in 2100, it accounts for 57 per cent, 54 per cent, 58 per cent, 47 per cent and 41 per cent of the world regions' total fuel supply to the respective transport sectors.

Ethanol and methanol become important particularly in WEU, EEU, LAM and PAS, where they account for 69 per cent, 88 per cent, 78 per cent and 86 per cent of the total transport supply, respectively. These two synthetic fuels will also be important for PAO (47 per cent), FSU (54 per cent) and SAS (52 per cent), together with other sources.

Oil, in the form of oil products, also remains as a source of supply, albeit with a limited role in NAM (14 per cent), LAM (14 per cent) and CPA (25 per cent). In other world regions, oil products account for only negligible shares of the total supply. The difference is due to differences in oil production, which strongly depends on the resource availability assumed for each region.

In the OG scenario, transport supply relies to a distinctly higher degree than PF on natural gas, with oil also providing continuous supply in some world regions. In PAS, MEA and CPA, compressed natural gas (CNG) vehicles account for 58 per cent, 49 per cent and 44 per cent, respectively of the total final energy supply to the transport sector in these world regions. In NAM, PAO and WEU, oil products still account for a large share of final energy in the transport sector: 46 per cent, 44 per cent and 33 per cent, respectively. In addition, in FSU, LAM, MEA and CPA, oil remains an important source of final energy in the transport sector, albeit with smaller shares.

Another notable feature of the OG strategy is the increasing importance of electricity in the transport sector. With the exceptions of LAM and PAS, electricity accounts for about 10 per cent of the total supply in each region, mainly in public railway transport.

5.2.2 Environmental Impacts

Different assumptions on technology strategies lead to different energy systems, which in general have different impacts on the environment. This is true in the near and the long term as well as on a world regional and the global level. We now proceed to presenting the environmental implications of choosing one or the other technology strategy. We focus on CO_2 and sulphur emissions as representing global and regional environmental impacts, respectively.

CO_2 emissions

The Kyoto Protocol (KP) was negotiated in December 1997 by the Third Conference of the Parties to the UN Framework Convention on Climate Change (FCCC). The KP includes six greenhouse gases (GHG), the emissions of which are to be limited in nearly all countries included in Annex I of the Convention; that is, the OECD countries (defined as in 1990) and countries in transition to market economies. The specified limits correspond to annual emissions of the six gases between 2008 and 2012, to about 95 per cent of their 1990 value (UNFCCC, 1997).

In the study presented in this book, we have restricted the analysis of GHG emissions to carbon dioxide (CO_2), the most important of the six gases covered by the Kyoto Protocol. Moreover, we cover only CO_2 emissions by the energy system and industrial sources (that is, emissions from fossil fuel combustion, gas flaring and cement production), and we have excluded greenhouse gas sinks from our analysis. These restrictions appear justified because we still cover the essential energy-related GHG emissions and, more importantly, their dependence on technology strategies.

The main source of anthropogenic CO_2 emissions is the combustion of fossil fuels. Worldwide, this led to the emission of 6.2 billion (10^9) tons of carbon (GtC) in 1990, of which 4.2 GtC were emitted in Annex I countries. Industrial processes, mainly cement production, accounted for another 0.2 GtC in the same year. Figure 5.26 describes the global CO_2 emissions from fossil fuel combustion and industrial sources between 1990 and 2100 for the two scenarios. CO_2 emissions in the OG strategy increase most of the time, reaching 31 GtC in 2100. This is still lower than the CO_2 emissions of 40 GtC that would be the result of carbon emissions growing as fast as the total energy demand during the same period.

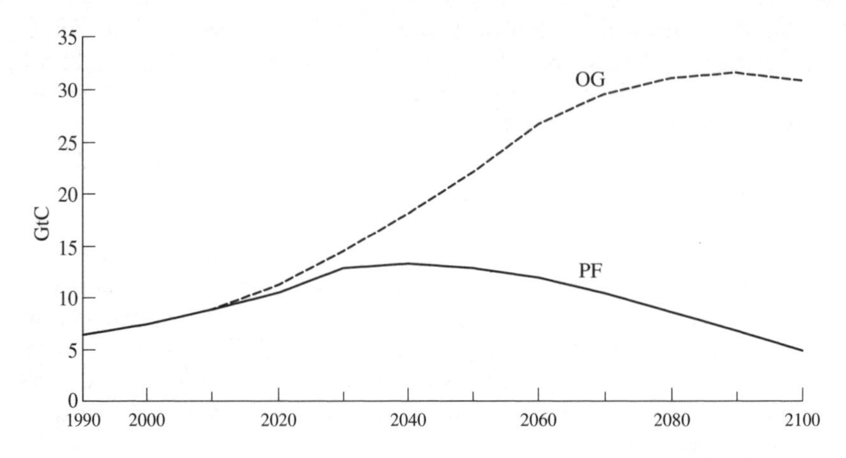

Figure 5.26 Global carbon emissions from fossil fuel combustion and industrial sources in two scenarios, 1990–2100 (GtC/yr), OG and PF strategies

In sharp contrast, PF shows CO_2 emissions by the end of the century that are lower than today's emissions. As a consequence of the inertia of the energy system, however, annual CO_2 emissions keep increasing even in PF until 2040; after that time, they begin to decline, reaching a level of 5 GtC in 2100. It is interesting to note that even the PF strategy, in which non-carbon technologies develop to a maximum extent, does not succeed in meeting the targets set by the Kyoto Protocol. If one agrees that greenhouse gas emissions in PF will lead to the goals of the UNFCCC being reached,[13] this scenario shows a possible way to meet the longer-term goals of the Framework Convention on Climate Change without meeting the Kyoto target. At the same time, as has often been noted, achieving only the Kyoto targets will not be sufficient to reach the goals of the UNFCCC.

Taking these two together, we argue that meeting the Kyoto targets is neither necessary nor sufficient for stabilizing the global climate. This may serve as a consolation in these days of a protocol that has lost some of its strength already. Our observation does not of course mean that the efforts to solve the climate change problems are small, but, for policy making, it has consequences for the timing of its solution. For research, it means that any analysis of climate change has to go significantly beyond the consideration of the Kyoto targets, and this was a major motivation for our study.

Figure 5.27 shows global carbon emissions in four world regions in the OG and PF scenarios. In the OG strategy, emissions from the OECD and REFS world regions increase little beyond the current level during the entire period. The increase of global carbon emissions is mainly due to the

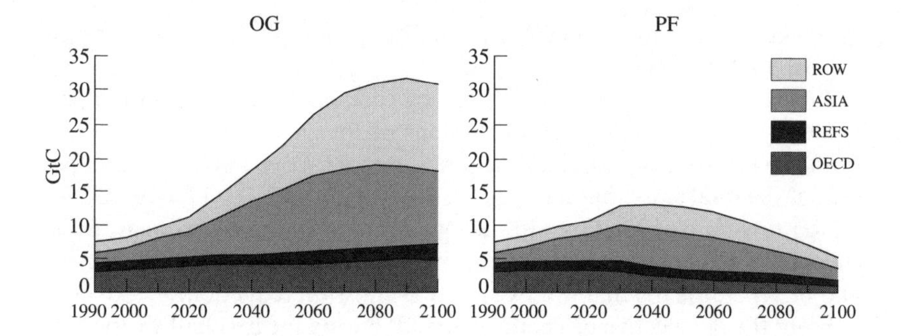

Figure 5.27 Anthropogenic CO_2 emissions by world region, 1990–2100, billion (10^9) tons of carbon (GtC), OG and PF strategies

increase in the Asian region, but also in the 'Rest of the World' (ROW) CO_2 emissions increase significantly, in particular between 2050 and 2100. In ROW, they increase, relative to 1990, by a factor of four by 2050 and by 2100 by a factor of as much as eight. Numerical values of world regional shares are given in Table 5.15.

For the OG scenario, the table shows that the 'Rest of the World' accounts for 42 per cent of the global carbon emissions in the OG in 2100. In the Asian region, the rapidly increasing emissions in the first half of the century (by a factor of 6.6) lead to as much as 45 per cent of the global emissions in 2050. After that year, Asian emission levels stabilize, however, but still account for roughly 35 per cent of the global emissions in 2100. In the same year the OECD region, which accounts for 41 per cent of global carbon emissions in 1990, reduces its share of the global total to 16 per cent.

The world regional shares of global carbon emissions for the PF strategy do not look very different from the OG picture. CO_2 emissions in the OECD and REFS regions are reduced to 27 per cent and 53 per cent of the current level, respectively, in 2100. In Asia, CO_2 emissions rise by a factor

Table 5.15 World regional shares of global CO_2 emissions (%)

	ASIA	OECD	REFS	ROW	Total
Actual (1990)	20	41	18	21	100
OG (2050)	45	19	7	29	100
OG (2100)	35	16	7	42	100
PF (2050)	43	17	10	31	100
PF (2100)	36	17	15	32	100

of 3.6 in 2050 but in 2100 will be reduced to a factor of 1.2, both in terms of year 1990 emissions.

Table 5.16 describes CO_2 emissions by economic sector. From the perspective of energy supply, CO_2 emissions of the energy supply and transformation sector (in particular the power sector) are responsible for 75 per cent of the total emissions increase between 1990 and 2100 (17.7 out of 23.6 billion tons of carbon, GtC) in the OG strategy. In the PF strategy, emissions from the energy supply sector will have decreased by the year 2100, and this sector is the main source of total emission reduction.

From the perspective of energy demand, during the first half of the 21st century, the main sources of the CO_2 emission increase in OG are from the residential and commercial (57 per cent of total increase) and the transportation (40 per cent) sectors, whereas for the second half of the century it is mainly transport (74 per cent of the total increase from 1990). In PF, CO_2 emissions from transport increase from 1.5 GtC to 2.3 GtC, but emission reductions in the residential and commercial and in the industry sectors lead to a reduction of the total.

Table 5.16 Global CO_2 emissions by economic sector for 1990, 2050 and 2100 (MtC), OG and PF strategies

	Actual	OG		PF	
	1990	2050	2100	2050	2100
Energy supply side					
Energy supply & transformation	2453	11 825	20 155	4976	1541
Direct use of fuels by sector	3782	9872	10 528	7623	3184
Non-energy emissions	1078	67	224	141	64
Total	7312	21 802	30 909	12 601	4789
Energy demand side					
Residential/ commercial	1995	10 290	3904	4846	1209
Industry	2784	4388	7813	2996	1244
Transport	1523	7262	18 967	4897	2334
Land use change	1010	−139	226	−139	2
Total	7312	21 802	30 909	12 601	4789

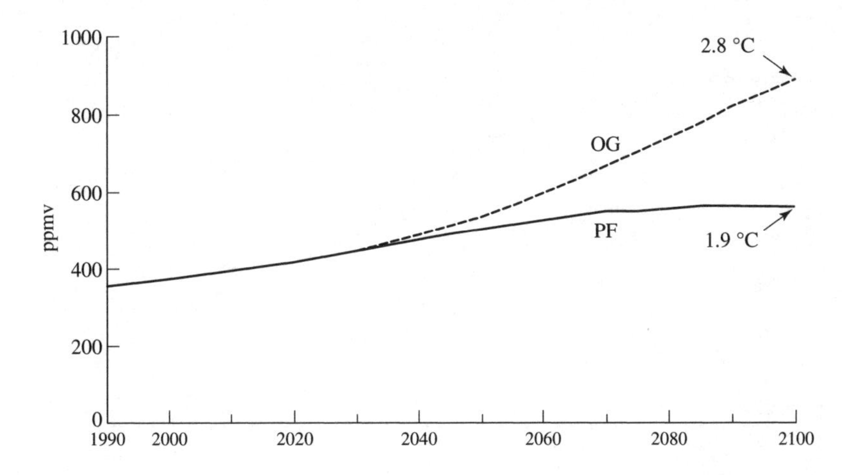

Figure 5.28 Atmospheric CO_2 concentrations in ppmv and central estimates of average global temperatures increases, OG and PF strategies

These emissions result in increases of atmospheric CO_2 concentrations from about 355 ppmv in 1990 to 900ppmv in 2100 for the OG strategy and to 560 ppmv for PF. Using IPCC's best guess (Houghton *et al.*, 1996) of climate sensitivity (that is, the global mean surface air temperature that accompanies a doubling of CO_2 equivalent concentrations), 2.5 Kelvin (K), surrounded by an uncertainty range from 1.5 to 4.5 K, OG's atmospheric CO_2 concentrations of 900 ppmv in 2100 result in a temperature rise of 2.8 K (middle value). The atmospheric CO_2 concentration of 560 ppmv in the PF scenario is estimated to result in a 1.9 K rise in the global mean temperature (Figure 5.28).

More importantly, Figure 5.28 shows that, in the PF scenario, atmospheric CO_2 concentrations have stabilized by 2100, whereas, in the OG scenario, they are still rising quite steeply.

Sulphur emissions

Sulphur oxides (SO_X) do not belong to the group of greenhouse gases in the narrow sense because they do not trap the heat radiation leaving the earth. Rather, they affect the atmospheric heat balance by 'shading', that is, by interrupting the flow of radiation energy travelling from the sun to the earth. This shading thus has a cooling effect and can therefore be regarded as beneficial for the climate because it counteracts global warming. However, SO_X has negative local and regional impacts on human health, food security and ecosystems.

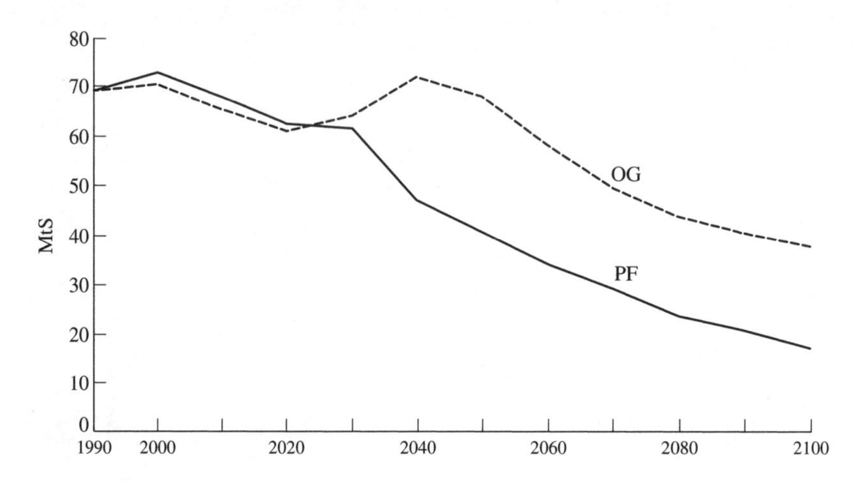

Figure 5.29 Global SO_X emissions, OG and PF strategies (MtS)

In 1990, the main sources of anthropogenic SO_X emissions were coal combustion (39 million tons of sulphur, MtS, and therefore more than half of the total) and oil combustion (17 MtS). Smaller emission volumes emanated from industrial activities (8 MtS), biofuel combustion (2 MtS) and international shipping (3 MtS). Total sulphur emissions calculated by the MESSAGE model for the two strategies are shown in Figure 5.29.

By the end of the 21st century, total sulphur emissions for the OG strategy are projected to have declined to about 37 MtS in 2100, from 69 MtS in 1990 (Figure 5.29). Along the way, SO_X emissions peak twice, once in the year 2000 and once in 2040. The first of these peaks reflects the peak of sulphur emissions in developed regions, and the second peak is mainly due to emissions in developing regions. It happens at a time when developed regions are already in the declining phase of their sulphur emissions.

Common to both of these phenomena is the assumption that increasing affluence leads to increased efforts to keep the local environment clean. Success of such efforts is depicted in what has been called 'the environmental Kuznets curve'. This curve shows a given environmental impact as the function of wealth (GDP per capita) and has the shape of an upside-down 'U'. The curve thus expresses decreasing pollutant emissions once an economy has reached a certain income level. (For a Kuznets curve on sulphur, see, for example, Grübler, 1998.)

For the PF strategy, SO_X emissions decrease throughout the model's time horizon, reaching, in 2100, a value as low as 25 per cent of their 1990 level.

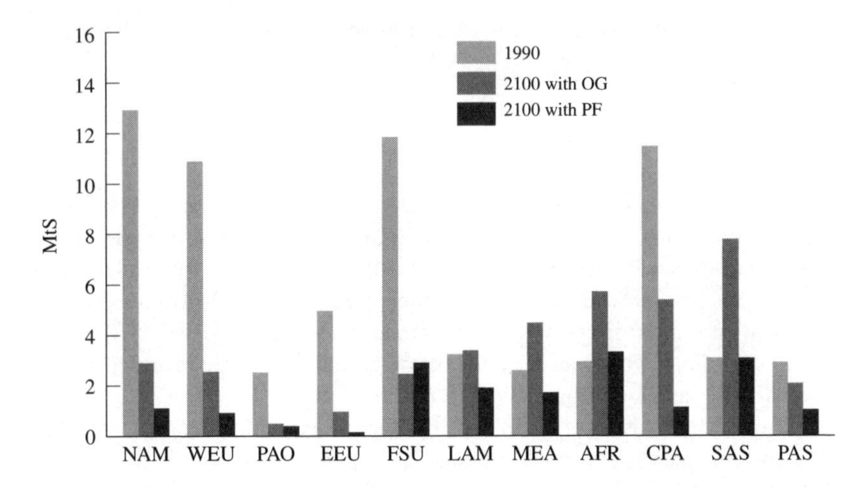

Figure 5.30 Global SO_X emissions, OG and PF strategies (MtS), 1990

Figure 5.30 shows the world-regional distribution of 1990 SO_X emissions (left), together with the 2100 values of the OG (middle), and the PF strategy (right). In 1990, four world regions, North America (NAM), Western Europe (WEU), the Former Soviet Union (FSU) and Central Planned Asia (CPA) emitted between 10 and 13 MtS, whereas most other regions emitted at relatively low levels (around 3 MtS), the EEU being an exception with 5 MtS. This picture changes completely by 2100.

With the OG strategy, SO_X emissions from NAM, WEU and FSU decline to around 2.5 MtS in 2100. By the same time, CPA reduces to around 5.5 MtS. LAM and PAS emit similar or slightly lower amounts of SO_X in 2100 to what they did in 1990. AFR and SAS increase their emissions relative to the base year. With almost 8 MtS of SO_X, SAS becomes the biggest SO_X emitter of all 11 world regions in 2100.

With the exception of the FSU, SO_X emissions are always smaller in PF than in OG. The reason for this exception is that the use of coal in district heat plants in this region is higher in PF than in OG owing to the delayed replacement of this technology by natural gas-fired technology. Compared with 1990, SO_X emission reductions in NAM, WEU and CPA are dramatic, that is, by an order of magnitude. Emissions in each of these three world regions decrease to around 1 million tons in 2100. Under the PF strategy, the EEU eliminates SO_X emissions almost completely, and also other regions emit less than they did in 1990.

5.2.3 Economic Implications

How much money do we need for the investments in the energy systems of the two scenarios? Figure 5.31 shows that, in 1990, total energy-related investment expenditures, which consist of investment for supply, electricity generation, production of synthetic fuels and off-grid electricity production (thus not including investments into end-use devices), were 570 billion (10^9) US dollars. This was 2.7 per cent of global economic product of that year, which was 21 trillion (10^{12}) US dollars.

In both scenarios, energy-related investment expenditures grow steadily, OG ending up with annual investment expenditures of 4.6 trillion US dollars, and PF with 2.8 trillion in 2100. The OG strategy thus turns out more costly than the PF strategy. In terms of (undiscounted) cumulative energy-related investment between 1990 and 2100, OG would cost 50 per cent more than PF (290 versus 190 trillion US dollars). As a percentage of GDP, however, both scenarios result in much lower shares of energy investments in world GDP in 2100 (550 trillion US dollars). Investment expenditures in OG are 0.8 per cent of that value in 2100 and 0.5 per cent in PF,

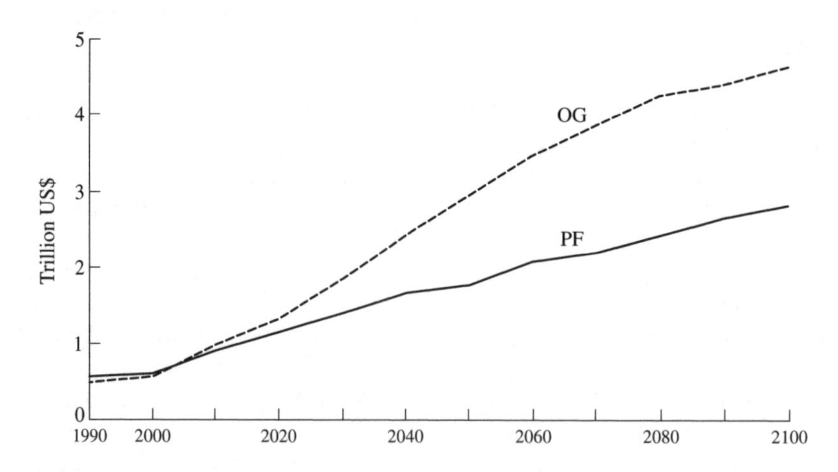

Figure 5.31 Energy-related investments, OG and PF strategies, in trillion (10^{12}) US dollars (constant 1990 US$)

compared with 2.7 per cent (570 billion out of a total GDP of 20.9 trillion) in 1990.

It would therefore be convenient if one could choose between OG and PF as from a restaurant menu. Most decision makers would then have no doubt in choosing the cheaper and environmentally more benign PF strategy. However, it is not that simple. One reason is that the economic and environmental advantages of the PF strategy come at a cost that is not explicitly included in the MESSAGE model. In our interpretation, it is the result of significant R&D efforts, which we attempt to quantify in the next section. We shall come back to this discussion in the next section (5.3).

Figure 5.32 shows the same annual investments into the energy systems in 1990, but this time disaggregated for the 11 world regions. Relative to all other world regions, North America (NAM) invested substantial amounts (34 per cent of the global total energy-related investments) in the energy sector. Other major investors were Western Europe (WEU), the former Soviet Union (FSU), Middle East and Northern Africa (MEA), and Pacific OECD (PAO).

Figure 5.33 shows the annual investments in the energy systems in 2100, by world region. In all 11 world regions, total investment cost in the year 2100 is higher with the OG strategy than with the PF strategy in all world regions. The strategy makes a particularly big difference in FSU, LAM, MEA, CPA and SAS. The difference is rather modest in currently developed regions (NAM, WEU and PAO).

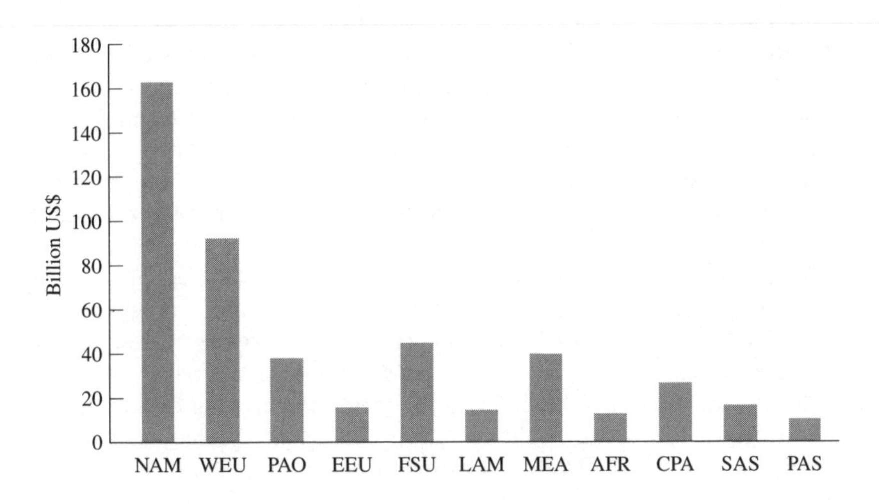

Figure 5.32 World regional energy sector investments, OG and PF strategies, billion (10^9) 1990 US$.

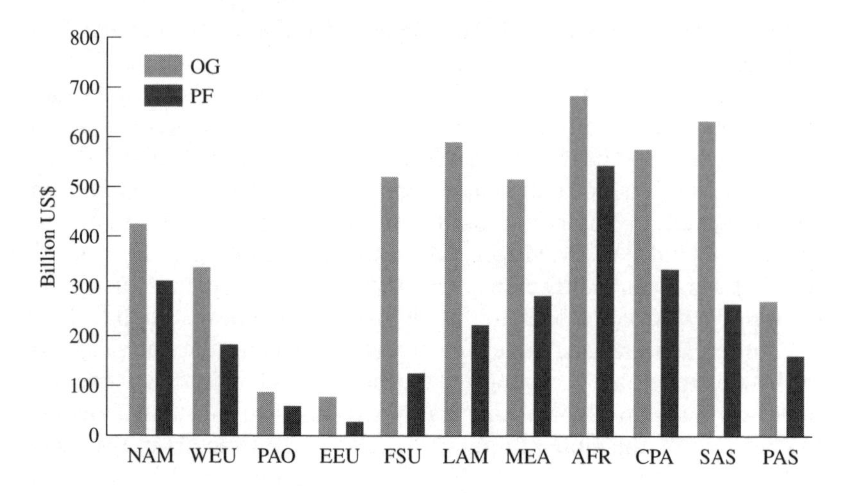

Figure 5.33 World regional energy sector investments in 2100, OG and PF strategies, billion (10^9) US dollars

5.3 ESTIMATION OF R&D EXPENDITURES PLUS POLICY IMPLICATIONS

Our preferred way of looking at the description of the differences between the OG and PF strategies is to regard them as illustrations of the way future energy systems can diverge as a result of pursuing one technology policy or another. By a policy that pursues the support of technological progress (assumed for the PF strategy) we primarily understand a policy that heavily supports research and development (R&D) of energy technologies. We therefore attempted a rough calculation of R&D costs and benefits.

We begin by showing the benefits of vigorous technological progress in the energy field by comparing undiscounted cumulative energy system costs for PF and OG in Figure 5.34. The figure shows that total energy systems costs in PF were 280 trillion US dollars less than in OG. In a very stylized way, then, this difference is the result of pursuing two strategically different technology policies, and it can be interpreted as the 'economic benefit' of successfully supporting technological progress in the PF strategy.

Let us now attempt an approximate calculation of R&D cost that might achieve the energy system cost reduction in PF relative to OG. For this

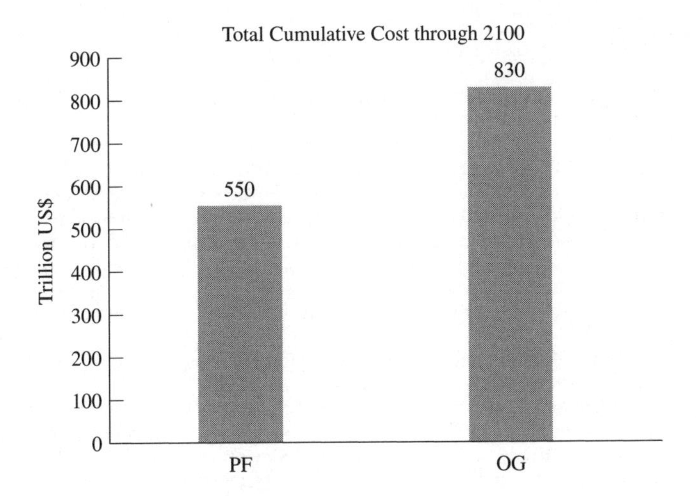

Figure 5.34 Total cumulative system cost through 2100 (undiscounted), OG and PF strategies, trillion (10^{12}) US dollars

calculation, we took results of the separate study on the R&D cost–benefit ratios calculated on the basis of the stylized ERIS (Energy Research and Investment Strategies) model (Miketa and Schrattenholzer, 2004). There the authors have found a ratio of 3.7 of R&D benefit to R&D expenditures on energy conversion technologies. Using this ratio and interpreting the difference of 280 trillion (10^{12}) US dollars as calculated from Figure 5.34 as R&D benefit, the required R&D expenditure to bring about this benefit is calculated as roughly 75 trillion US dollars. Owing to the very preliminary nature of the results reported in the publication quoted, this number must be taken only as a very rough indicator. In our opinion, this should be taken as a representative value of an interval at least as wide as ranging from 50 to 200 trillion.

This back-of-the envelope calculation ignores the effect of discounting. Adding discounting would reduce the benefit–cost ratio because, obviously, research and development has to occur before its benefits can be reaped. Maybe the bigger factor explaining why the world had not visibly embarked on this road is the general uncertainty of pay-off to investments, but we also think that it could be a lack of imagining the benefits of technological progress, not only in terms of monetary payback, but also in terms of the environmental compatibility of the global energy system. In this regard, we hope that this study contributes to 'whetting the appetite' of policy makers for technological progress and its support by policy.

NOTES

1. Our OG strategies scenario corresponds to the A1G scenario in the IPCC report, and the PF strategies scenario corresponds to the A1T scenario.
2. The four 'macro' world regions and their definition in terms of the 11 'IIASA world regions' are OECD (North America, Western Europe and Pacific OECD), Reforming Economies (REFs: Eastern Europe and Former Soviet Union), Asia (China, South Asia and Other Pacific Asia) and Rest of the World (Latin America, Sub-Saharan Africa, and Middle East and Northern Africa). Note that 'Rest of the World' (ROW) is dubbed ALM (Africa and Latin America) in the SRES. For a world map with the definition of the 11 regions, see the appendix.
3. The SRES-B1 scenario, briefly characterized in section 3.3.3, is more of the 'small is beautiful' type.
4. For a more detailed discussion of the issue, readers are referred to IPCC's Special Report on Emission Scenarios (SRES; Nakićenović *et al.*, 2000).
5. The seven demand sectors are industry-specific (electricity and electricity substitutes), industry non-specific (including thermal energy), residential/commercial specific (electricity and electricity substitutes), residential/commercial non-specific (including thermal energy), transport, non-commercial and feedstocks.
6. It is reasonable to assume that this interplay is non-linear. To endogenize non-linear relations for 400 energy technologies would result in a model that would be unsolvable under practical circumstances.
7. For this reason, it might be better to use the more general term 'methane' rather than 'natural gas' as the term to describe this resource category.
8. Typical of emerging technologies is that they are allowed to double their installed capacities within ten years.
9. The primary-energy equivalent of nuclear electricity and electricity generated from renewable energy (including hydropower) was calculated using the so-called 'direct equivalent method'. This means that the primary-energy equivalent of electricity generated by these sources is calculated as if no energy would be lost during the conversion from primary to final energy. This method avoids calculating unrealistically high primary-energy values, in particular for solar energy, which would be the result of counting the radiation energy received from the sun as primary energy instead.
10. These technologies produce off-grid electricity. Examples of this category are solar panels, hydrogen mini-turbines and by fuel cells.
11. This corresponds approximately to three times the total global primary-energy consumption in 1990.
12. Note however, that FSU is still the biggest producer of natural gas among all world regions. The reasons for the small share of gas-based electricity generation in FSU are, first, that 70 per cent of the gas produced is exported to other world regions, and second, that gas consumption in FSU is predominantly used directly in the industry sector, rather than for power generation, where nuclear energy dominates.
13. The atmospheric concentration of greenhouse gases in the PF scenario is estimated to stabilize at 560ppmv (see below). It is as yet unknown whether this will avoid 'dangerous interference' with the global climate system as stipulated in the UNFCCC.

REFERENCES

Barro, R.J. (1997), *Determinants of Economic Growth*, Cambridge, MA: MIT Press.

Gritsevskii, A. (1996), 'Scenario Generator', Internal Report, International Institute for Applied Systems Analysis (IIASA), Laxenburg, Austria.

Grübler, A. (1998), *Technology and Global Change*, Cambridge: Cambridge University Press.

Houghton, J.T., L.G. Meira Filho, B.A. Callander, N. Harris, A. Kattenberg and K. Maskell (eds) (1996), *Climate Change 1995. The Science of Climate Change*, contribution of Working Group I to the Second Assessment Report of the Intergovernmental Panel on Climate Change, Cambridge: Cambridge University Press.

Lutz, W., W. Sanderson and S. Scherbov (1997), 'Doubling of world population unlikely', *Nature*, **387**(6635), 803–5.

Lutz, W., W. Sanderson, S. Scherbov and A. Goujon (1996), 'World population scenarios in the 21st century', in W. Lutz (ed.), *The Future Population of the World: What Can We Assume Today?*, 2nd rev. edn, London: Earthscan, pp.361–96.

Maddison, A. (1995), *Monitoring the World Economy 1820–1992*, OECD Development Centre Studies, Paris: Organization for Economic Co-operation and Development.

Masters, C.D., E.D. Attanasi and D.H. Root (1994), *World petroleum assessment and analysis*, Proceedings of the 14th World Petroleum Congress, Chichester, UK: John Wiley & Sons.

Miketa, A. and L. Schrattenholzer (2004), 'Experiments with a methodology to model the role of R&D expenditures in energy technology learning processes; first results', *Energy Policy*, **32**(15), 1679–92.

Nakićenović, N. and R. Swart (eds) (2000), *Emissions Scenarios, Special Report of the Intergovernmental Panel on Climate Change*, Cambridge: Cambridge University Press.

Nakićenović, N., A. Grübler and A. McDonald (eds) (1998), *Global Energy Perspectives*, Cambridge: Cambridge University Press.

Nakićenović, N., A. Grübler, H. Ishitani, T. Johansson, G. Marland, J.R. Moreira and H.-H. Rogner (1996), 'Energy primer', in R. Watson, M.C. Zinyowera and R. Moss (eds), *Climate Change 1995. Impacts, Adaptations and Mitigation of Climate Change: Scientific Analyses*, Cambridge: Cambridge University Press, pp.75–92.

Rogner, H.-H. (1997), 'An assessment of world hydrocarbon resources', *Annual Review of Energy and the Environment*, **22**, 217–62.

Rogner, H.-H., B. Fritz, M. Gabrera, A. Faaij and M. Giroux (2000), 'Energy resources', in J. Goldemberg (ed.), *World Energy Assessment Report*, New York: UNDP, UNDESA, WEC.

Strubegger, M., A. McDonald, A. Gritsevskii and L. Schrattenholzer (1999), *CO2DB Manual, Version 2.0*, April, IIASA, Laxenburg, Austria.

UNDP (United Nations Development Program) (1993), *Human Development Report 1993*, New York: Oxford University Press.

UNFCCC (United Nations Framework Convention on Climate Change) (1997), *Kyoto Protocol to the United Nations Framework Convention on Climate Change*, FCCC/CP/L7/Add.1, 10 December 1997, New York: UN.

6. Summary and policy implications

Can energy–economy–environment (E3) scenarios that reach 100 years into the future be policy-relevant? And, if so, what kind of guidance can they provide to today's policy making? In general, it should be clear that the policy relevance of long-term scenarios has to be different from that of near-term outlooks, the difference coming from different objectives. Typical near-term objectives with respect to the E3 system are, for instance, economic viability and immediate environmental impact of given project alternatives. In contrast, examples of long-term objectives would be economic, social and environmental sustainability, the latter including climate protection.

From this difference it follows that policy relevance of near-term outlooks tends to be concrete – for instance, in assessing payback times and the environmental impact of alternative project variants – whereas the policy relevance of long-term scenarios is more strategic. To emphasize this aspect, we have, in some places, used the term 'strategy' to refer to a scenario. In the E3 system, and in other fields too, strategies aim at achieving a definite favourable outcome, which, in our case, is the sustainable development of the global energy–economy–environment system.

The timing of strategic policies is a delicate issue. Although the target of sustainable development may be far away, this does not mean that there is much time to wait. We have used the term 'slow variables' to refer to the driving forces that are central to the understanding of the long-term developments analysed in this book. The term refers to a general inertia of these variables, which makes their projection more reliable, but which at the same time means that a policy aiming at influencing their development must consider the possibility of long lead times and therefore try to avoid significant delays.

For sustainable development at large, quantification is an important step towards making the concept operational for policy making. Accordingly, a large number of sustainability indicators have been proposed in the relevant literature. However, there is no generally agreed-upon indicator or set of indicators, let alone threshold levels that must be reached in order to achieve sustainability. The reason for this lack of precise criteria is, in our opinion, the generality and multidimensionality of the sustainability concept, which makes an unambiguous quantification difficult. The most important source of difficulties is that different objectives are often in conflict with each other. Another difficulty is that some of the proposed indicators are difficult to measure.

To increase the operational aspect of sustainability despite these conceptual difficulties, we introduced a definition of sustainable-development E3 scenarios in Chapter 1 and used it to characterize E3 scenarios as either sustainable or non-sustainable. While maintaining the spirit of the well-known and generally accepted 'Brundtland definition', according to which sustainable development (SD) meets the needs of the present without compromising the ability of future generations to meet their own needs, we have restricted the definition of SD scenarios to a compact set of four criteria, covering economic and environmental sustainability as well as interregional (that is, intragenerational) and intergenerational equity. Our definition thus also addresses the three 'pillars' of economic, environmental and social sustainability.

By definition, sustainable-development E3 scenarios describe the E3 system within the chosen boundaries. Issues like water supply, agriculture and biodiversity, for example, have not been addressed here and, in that sense, we have presented 'partial' sustainable development. However, with a focus on the energy system, we claim that we have captured the essence of sustainable development of the global energy system in an adequate way. This is to say that we think that the sustainability within the boundaries chosen for our analysis can be a solid building block for the design of more general SD strategies.

Another point to note about our definition of sustainable-development E3 scenarios is that it is an ex post definition; that is, it classifies existing scenarios. This is probably not a major point, but it indicates that none of our scenarios was constructed in a deductive way 'from first principles'. This may explain one or the other feature to some readers, but, more importantly, we want to emphasize that our criteria are like a scale that measures the extent to which already existing descriptions of the future E3 system are compatible with sustainable development. They do not directly provide an operational instruction on the way to design sustainable development, but they provide a frame for aiming policies at achieving sustainability in the long run.

In our opinion, using ex post criteria is an advantage because it means that assumptions were not biased in a way to lead to scenarios that qualify as SD scenarios. At the same time this means that any single SD scenario is sufficient but not necessary (in a formal logical sense) for the sustainable development of the global energy system, and the existence of other SD scenarios is obvious. Still, talking in a more colloquial sense, we think that our SD scenarios also give good indications of what kind of action will be necessary to achieve sustainable development of the energy system and that their usefulness for policy making comes from providing typical illustrations of a future sustainable development of the global energy system.

Following the presentation and discussion of our criteria for sustainable-development scenarios in Chapter 1, we defined the term 'scenario' and explained the methodology of generating them in Chapter 2. On this basis, we presented, in Chapter 3, a selection of IIASA E3 scenarios, classifying them into three groups: high-impact, carbon mitigation and sustainable-development scenarios.[1] As a background and reference, we used the database of long-term E3 scenarios as compiled during the work on the *Special Report on Emission Scenarios* (Nakićenović and Swart, 2000) of the Intergovernmental Panel on Climate Change (IPCC). We could show that the 34 IIASA global scenarios selected for presentation in this book were, in terms of the most important variables describing the evolvement of the global E3 system, representative of all the scenarios in the SRES database.

As a basis for specifically discussing the policy relevance of the sustainable-development scenarios, we first identified the typical features of these scenarios in the following areas: population growth, economic growth, the interregional wealth gap and decarbonization. Relative to all scenarios collected in the IPCC-SRES database, SD scenarios are characterized by relatively low population, high economic growth, a significant narrowing of the income gap, and faster decrease in energy and carbon intensities. For policy making, these characteristics suggest the following. Policies in today's developing regions (as is well known, many industrialized countries face the opposite problem of too low birth rates) should be aimed at stabilizing total population during the course of this century. In order to narrow the income gap between industrialized and developing regions, policies in industrialized regions should aim at facilitating economic growth in developing regions. Doing so may increase industrialized regions' expenditures in the short run, but achieve sustainable development in the long run.

Policies to achieve decarbonization at rates characteristic of SD scenarios would be policies that aim at technological progress, not only of carbon-free technologies such as those utilizing renewable energy, but also of efficient energy conversion technologies. Whereas carbon-free technologies directly reduce the carbon intensity of energy conversion and use, the improvement of conversion efficiency leads to a reduction of the carbon intensity of economic output by reducing its energy intensity. Outside the technological domain, energy intensity of GDP can also be achieved by policies aiming at the reduction of final energy required for a given energy service. This last step of the energy chain has the biggest potential for saving energy, and examples for such policies abound. Let us just mention policies to enhance the attractiveness of public transport or car pooling and the introduction of standards such as the minimum thermal efficiency of buildings or the energy consumption of appliances.

The IIASA-ECS scenarios address technological progress of energy conversion by defining the so-called 'reference energy system' in a degree of detail that includes approximately 400 energy conversion technologies from primary-energy extraction to end use. In order to organize the presentation of more detailed results on technological change in E3 scenarios, we introduced the concept of technological clusters in Chapter 4 to enable us to speak efficiently about groups of technologies. This cluster concept, originally meant to categorize similar technologies by technological criteria, turned out also to be well suited to characterizing scenarios. Clusters that characterize scenarios and groups of scenarios are defined, not so much by technological similarity, but by public acceptance (defined in a way to be consistent with the general spirit of a scenario as captured in the scenario's 'storyline') and by 'market success'. The high public acceptance and the market success of solar energy and hydrogen as a final energy carrier was the most robust result of the analysis of SD scenarios.

Having thus identified patterns of the evolvement of important systems variables and the market success of solar and hydrogen technologies in SD scenarios, we then presented a more detailed analysis of the post-fossil (PF) scenario as an illustration of a typical SD scenario in Chapter 5. To substantiate the characteristics of this SD scenario, we described an oil and gas-rich (OG) scenario, which turned out to be a non-sustainable scenario, in parallel to the PF scenario. The difference between the two scenarios is predominantly in the technological sphere. In the OG scenario, technologies related to the production and use of oil and natural gas are assumed to make particularly fast progress, whereas, in the PF scenario, technologies harvesting and converting renewable energy are assumed to progress faster. The two scenarios are identical for other parts of their respective 'storylines', in particular with respect to economic growth in all world regions.

In the spirit of scenario analysis, OG and PF represent two out of many possible development paths of the global E3 system. In order to derive policy implications from these two scenarios, let us, for a moment, assume that these two scenarios describe the only two possible developments and discuss the mechanisms that will lead to the realization of one versus the other. Broadly speaking, the outcome will be determined by a combination of random events that are beyond our control and decisions on the course of events that are under our control.[2]

In our energy model, decisions on events that are under our control are subject to optimization, and uncertain events are reflected by model input parameters. Different states of the world can either be specified in advance as input parameters or included (endogenized) as model variables. The choice between these two possibilities depends, among others, on the numerical feasibility of solving large and nonlinear models and on the existence of theories that allow the formulation of mathematical relationships between the model variables. More often than not, this choice is not clear-cut, and both possibilities appear attractive.

In this study, such a borderline case concerned the influence of policies on technological progress. In general, we assume that research and development (R&D) can tilt the technological developments into one or the other direction. In the two scenarios that we analysed in detail, this relationship is reflected in two different storylines. The storyline of the OG (oil and gas-rich) scenario specifies that technological research and development is successfully aimed at the development of oil and gas technologies and that of the PF (post-fossil) scenario makes the same specification for the development of non-fossil technologies. Thus what has been modelled in our study as two different states of the world (that is, the success of two different kinds of technologies) could also have been modelled as the result of specific R&D efforts.

To the extent that developments in the PF scenario are the result of specific R&D, the detailed description of this scenario can be interpreted as a 'roadmap' to sustainable development. In the real world, however, such a result of supporting technological progress cannot be expected with certainty, but the prospect provided by the mere portrait of a possible SD scenario is, in our opinion, a prerequisite for the realization of the strategic goal of sustainable development, and the trajectories of this scenario can serve as milestones.

In our model, technological progress is quantified in several different ways. Progress of primary-energy extraction is reflected by increasing amounts of fossil primary-energy reserves. Progress of energy conversion technologies is modelled as specific technology costs decreasing and conversion efficiencies increasing over time. Optimistic as these cost reductions may appear, they are just the result of extrapolating a surprisingly regular trend that has been observed in the past. Take solar photovoltaic (PV) energy conversion as an example. It has been observed that specific costs of PV modules have declined by more than 20 per cent for each doubling of cumulative production of this technology. This regularity persisted over more than one order of magnitude of PV cost reduction in the past.

Such regular cost reductions are the main feature of the so-called 'learning' (or experience) curves, which reflect the empirical observation that specific technology costs decline at constant (learning) rates for each doubling of cumulative installed capacity of a given technology. For a more detailed description of this concept and a survey of learning rates of energy technologies, see McDonald and Schrattenholzer (2002). The policy relevance of experience curves comes from the fact that, for the society as a whole, it may be cheaper in the long run to deploy technologies at a time when they are not yet competitive. To ensure that this overall optimum is reached, incentives (for instance, subsidies) must be provided that make the more costly technology economically attractive to consumers. Another possibility to accelerate the 'learning' of a technology could be its public procurement.

At this point, some readers may miss the discussion of carbon taxes as a policy instrument to promote the introduction of carbon-free energy. To some extent, this is a technical point because in carbon mitigation scenarios, for instance, carbon taxes are included implicitly as a consequence of the carbon constraint. More important is the lack of carbon taxes in the SD scenarios such as in the PF scenario. In those scenarios, the respective storylines assume a general preference in society for environmentally compatible energy supply. Public and private research efforts together with consumer preferences reinforce each other, thus leading in the end, to a clean energy system in which renewable energy can be sold at competitive prices.

Throughout this book, we have taken due account of the uncertainty that is inherent in the presentation of developments that reach as far into the future as our analysis. However, a sensitivity analysis of long-term E3 scenarios can never be complete. At the end of the main part of this book, we would therefore like briefly to sketch sensitivities and alternative scenarios.

Beginning with the SRES scenarios, an SD scenario that is distinctly different from the PF scenario is the B1 scenario, which we have briefly sketched in Chapter 3. Broadly characterized, B1 is more of the 'small is beautiful' type than PF. Whereas, in PF, economic growth guides technological progress and environmental compatibility of energy technologies, in B1 environmental consciousness is the main driving force towards sustainability. In policy terms, sustainable development in B1 is the result of decentralization and 'dematerialization', that is, a high degree of economic activity in the less energy-intensive service sector. Perhaps one could therefore say that B1 is thus 'greener' than PF, which relies more on technological progress to solve problems of unsustainable development.

On the primary-energy side, dropping the assumption (made in the PF scenario) that sustainable development relies to a major extent on renewable energy could lead, for example, to an SD scenario that includes massive shares of nuclear energy. In such a scenario, hydrogen could be generated by thermal water splitting in high-temperature reactors, a scheme proposed some 30 years ago by Marchetti (1976). But also coal could be the dominant primary-energy carrier of an SD scenario if 'clean coal' technologies, including carbon sequestration and storage, are deployed. Current work at IIASA studies such options in greater detail. In all alternative scenarios briefly mentioned here, hydrogen technology plays a major role. It would therefore appear as a robust conclusion for technology policy that hydrogen technology, in particular fuel cells, is a key to sustainable development.

In connection with uncertainty, readers may ask us about our assessment of the likelihood of any of the SD scenarios actually materializing. We address this question here despite having no precise answer. So let us approach this question one step at a time. In response to policy makers, one part of the answer is that we have written this book with the intention of providing images of possible sustainable future developments of the global energy system (and the conditions under which they might materialize) for the purpose of making it attractive for policy makers to contribute to the establishment of an environment in which sustainable development could prosper. Policy makers can therefore get an idea of the probability of sustainable development by assessing their own probability that they might do their best to create this environment.

For readers without influence on policy making, the same general idea still applies, albeit to a much lesser degree. What they could do to enhance the chances of sustainable development would be to internalize the spirit of the SD scenarios presented here. As we have argued in our cluster analysis in Chapter 4, the public acceptance of policy goals is a crucial part of the 'storyline' defining any particular E3 scenario. As a further contribution to the question of likelihood of SD scenarios, we shall now briefly discuss under which circumstances they might be realistic.

We agree with de Vries *et al.* (2000) that the likelihood of such SD scenarios actually materializing depends largely on whether the trends towards globalization and worldwide cooperation can be accelerated. In addition to obstacles coming from possible free-riding behaviour and from shortsighted preference of near-term consumption over investments in long-term sustainability, the drive towards cultural identity and diversity might be strongly working against globalization and liberalization (see, for example, Huntington, 1997).

Another way of approaching the likelihood question would be to ask whether the efforts to achieve sustainable development would be large or small. Of course, no clear answer can be given, because even if we could quantify a response, surely there would be no agreement as to whether this number is big or small. Still, we want to approach an answer by asking related questions. How big were the efforts made to land humans on the moon? How expensive is it to prepare military defence against actual or imagined threats? How expensive would it be today to accommodate an oil price of 100 US dollars per barrel (as projected in the early 1980s)? Of course, the additional questions cannot be answered either. We mention them here to illustrate scenarios in which substantial efforts are being made or would be made in a realistic setting. We cannot guarantee that sustainable development would be achievable with comparable efforts, but any of these efforts implied in our examples would carry the world a long way towards achieving it. Lack of attractiveness of the goal should be no obstacle.

NOTES

1. As we have explained in Chapter 3, although most of the mitigation scenarios reported here do not meet the sustainable-development criteria, the distinction between mitigation scenarios and SD scenarios is not totally clear-cut. In order to emphasize that sustainable development is a more general goal than climate mitigation, we have classified (the few) mitigation scenarios that qualify as SD scenarios simply as mitigation scenarios.
2. A general example of an event beyond our control is the success of research and development activities. An example of decisions on events that are under our control is a decision to invest in a specific piece of energy conversion technology.

REFERENCES

de Vries, B., J. Bollen, L. Bouwman, M. den Elzen, M. Janssen and E. Kreileman (2000), 'Greenhouse-gas emissions in an equity-, environment- and service-oriented world: an IMAGE-based scenario for the next century', *Technological Forecasting and Social Change*, **63**(2–3).

Huntington, S. (1997), *The Clash of Civilizations and the Remaking of World Order*, New York: Simon & Schuster.

Marchetti, C. (1976), 'Geoengineering and the energy island', in W. Häfele *et al.* (eds), *Second Status Report of the IIASA Project on Energy Systems*, RR-76-1, Laxenburg, Austria.

McDonald, A. and L. Schrattenholzer (2002), 'Learning curves and technology assessment', *International Journal of Technology Management*, **23**(7/8), 718–45.

Nakićenović, N. and R. Swart (eds) (2000), *Emissions Scenarios, Special Report of the Intergovernmental Panel on Climate Change*, Cambridge: Cambridge University Press.

Nakićenović, N., J. Alcamo, G. Davies, B. de Vries, J. Fenhann, S. Gaffin, K. Gregory, A. Gruebler, T.Y. Jung, T. Kram, E.L. La Rovere, L. Michaelis, S. Mori, T. Morita, W. Pepper, H. Pitcher, L. Price, K. Riahi, R.A. Roehrl, H.-H. Rogner, A. Sankovski, M. Schlesinger, P. Shukla, S. Smith, R. Swart, S. van Rooijen, N. Victor and Z. Dadi (2000), *Special Report on Emissions Scenarios* (SRES), A Special Report of Working Group III of the Intergovernmental Panel on Climate Change, Cambridge: Cambridge University Press.

Appendix MESSAGE: a technical model description

Manfred Strubegger, Gerhard Totschnig and Bing Zhu

1. INTRODUCTION

The main tool used for the generation of the scenario results presented in this book is the MESSAGE model. For many readers, the main part of the book will provide an adequate description of MESSAGE and its input data. However, more technically oriented readers may be interested in seeing a more detailed model specification, which we therefore present in this appendix.

MESSAGE (Model of Energy Supply Systems Alternatives and their General Environmental Impacts) is a systems engineering optimization model used for medium to long-term energy planning, energy policy analysis and scenario development. The roots of its development go back to IIASA's Energy Systems Program of the 1970s. MESSAGE has been used for many projects and scientific studies. More recent examples of these include the joint IIASA-WEC report on Global Energy Perspectives (Nakićenović *et al.*, 1998), the IPCC Special Report on Emissions Scenarios (Nakićenović *et al.*, 2000), and the IPCC Third Assessment Report (Metz *et al.*, 2001).[1] The most recent version of the model is known as MESSAGE V.

MESSAGE finds the optimal flow of energy from primary energy resources to useful energy demands, which is feasible in a mathematical and an engineering sense, and at the same time represents the investment choices that lead to the lowest cost of all feasible energy supply mixes to meet the given energy demand. Engineering feasibility is ensured by making energy flows consistent with model constraints on primary-energy extraction, energy conversion and transport as well as on end-use technologies. Such energy flows are further determined by constraints on the rate of new capacity installation (new capacity can be installed only gradually), the substitutability among energy forms, resource recoverability, renewable-energy potentials and others. Such flows are determined for each geographical world region as specified below, page 174.

Out of many possible energy flows, MESSAGE selects the one that supplies the exogenously given demand *at least cost*. The optimization process thus can be likened to decision makers who invest in energy technologies characterized by different performance, cost and environmental characteristics in such a way as to meet demands at least cost under the given constraints. Changes in the energy system are therefore endogenous, that is, the pace of enhanced energy conservation or a structural change is determined by shifts in the technological applications selected. Costs include investment costs, operation and maintenance costs (fixed and variable), fuel costs and any user-defined costs such as environmental costs of pollution. Calculating total cost, MESSAGE uses assumptions on specific costs of hundreds of individual technologies as they develop over time. The actual number of technologies is specified in the MESSAGE input files. The function used to calculate the costs is called 'objective function'. For determining cost optimality in the scenarios presented in this book, all costs are discounted using an annual rate of 5 per cent.

The result of these two steps (establishing feasibility and then calculating the optimal supply path) is an optimal energy supply mix by different energy supply technologies and different energy carriers.

In the description here, we attempt to focus on the model as it was used for the scenario runs included in this book, but in some cases we also add brief descriptions of model features that have not been used for the reported runs. We think that a model description would be incomplete without these additions. Such features not used in the scenarios of the book are the mixed-integer option, supply and demand elasticities, load regions, storage variables and the capability of the model to do multi-objective optimization.

The Reference Energy System

The conceptual core of the MESSAGE model is the Reference Energy System (RES). It provides the framework for representing an energy system with all its interdependencies from resource extraction, imports and exports, conversion, transport and distribution to the demand for energy end-use services (that is, useful-energy demand). Useful energy provides the consumer with energy services such as cooking, illumination, space conditioning, refrigerated storage, transport, industrial production processes and consumer goods. The purpose of the energy system is, therefore, the fulfilment of demand for energy services. A schematic illustration of the RES is given in Figure A.1.

The RES includes major energy flows and conversion technologies that are available to the model in 1990, and those that play an increasing role in a number of scenarios during the remainder of the model's time horizon,

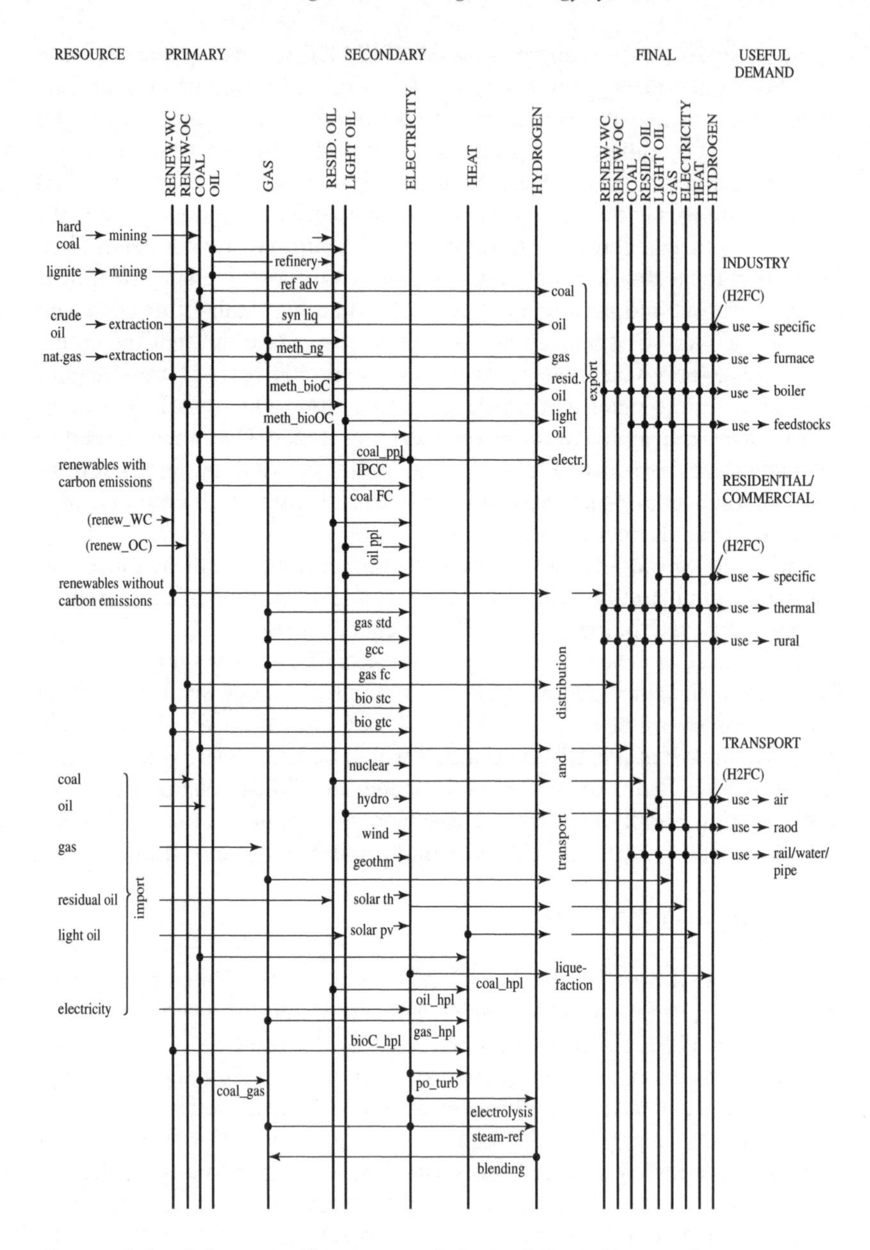

Figure A.1 Schematic illustration of the Reference Energy System

Key:

ref_adv	Advanced refinery
meth_ng	Methanol production from natural gas

the year 2100. Note that Figure A.1 gives just a schematic overview of the RES. The full energy system of the MESSAGE model includes various additional energy carriers and conversion technologies (for example, the full nuclear cycle including reprocessing of nuclear fuels).

In this appendix, the term 'energy conversion technology' refers to all kinds of energy technologies from resource extraction to transformation, transport and distribution of energy carriers and end-use technologies. Through the definitions of energy carriers and technologies so-called 'energy chains' are defined: all possible (feasible) energy flows from resource extraction or imports to the useful-energy demand. The demands, which are inputs to the model, must be met by the energy supplied through one or more of the modelled energy chains.

Since few energy conversion technologies convert resources directly into useful energy, intermediate *energy levels* (secondary, final) can be defined in the scenarios. Each flow between two nodes in Figure A.1 represents an energy conversion technology linking different energy levels. The interaction of energy conversion technologies is determined in the RES by the specification of the energy levels from which they take their inputs and to which they deliver their outputs. Figure A.2 gives an example of an energy chain and of the energy levels used in many

Figure A.1 (*continued*)

meth_bioC	Methanol production from biomass with carbon emissions
meth_bioC0	Methanol production from biomass without carbon emissions
syn_liq	Coal liquefaction and synthesis of light oil
coal_ppl	Coal power plant
IGCC	Integrated gasification combined cycle plant
coal FC	Coal fired fuel cell
oil ppl	Oil-fired power plant
gas std	Natural-gas power plant (single steam cycle)
gcc	Natural-gas power plant (combined cycle)
gas fc	Natural-gas fired fuel cell
bio stc	biomass-based gasification power plant (single steam cycle)
bio gtc	biomass-based gasification power plant (combined cycle)
solar th	Solar thermal power plant
solar pv	Solar photovoltaic power plant
coal_hpl	Coal-fired heating plant
oil_hpl	Oil-fired heating plant
gas_hpl	Natural gas-fired heating plant
bioC_hpl	Biomass heating plant with carbon emissions (non-sustainable use of biomass)
po_turb.	Pass-out turbine
steam-ref	Steam reforming of natural gas to hydrogen
resid.oil	Residual Oil
Renew-WC	Renewables with carbon emissions (non-sustainable use of biomass)
Renew-0C	Renewables without carbon emissions (sustainable use of biomass)
H2FC	Hydrogen fuel cell
nat. gas	Natural Gas

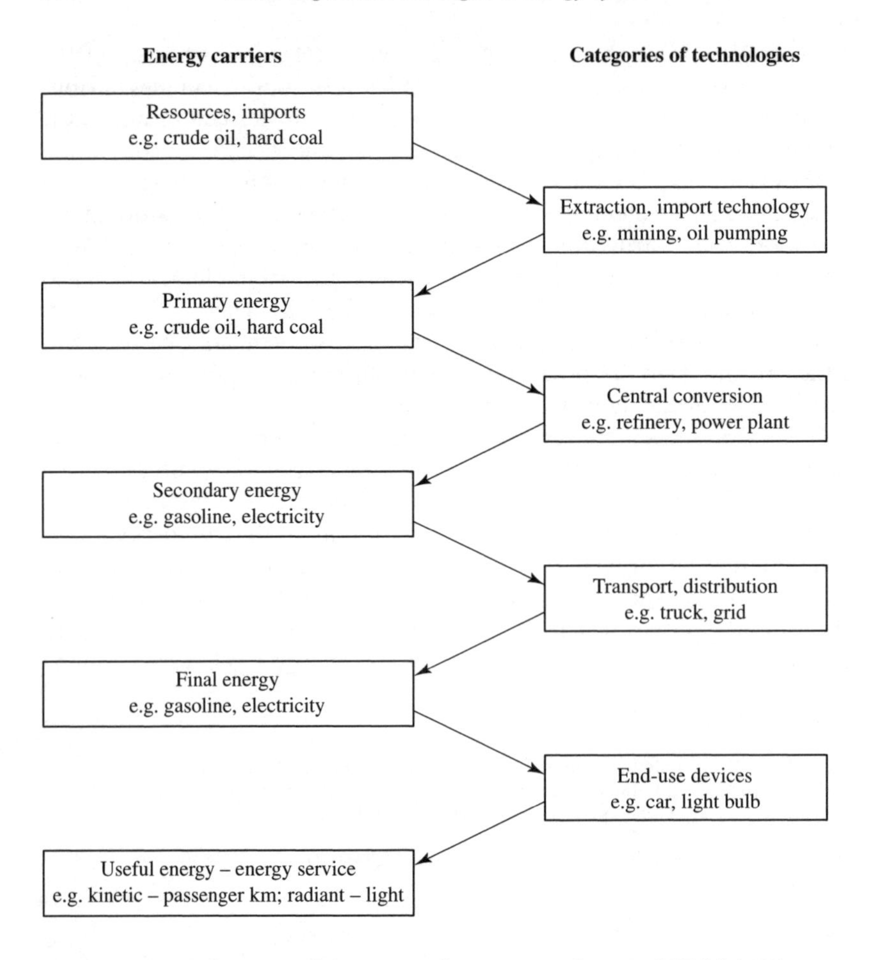

Figure A.2 Schematic illustration of an energy chain in MESSAGE

scenarios, together with the associated energy carriers and categories of conversion technologies.

For building the names of model variables and constraints, the following characters are used to identify the energy levels:

R = energy resources,
A = primary energy,
X = secondary energy,
T = final energy (after transmission),
F = final energy (after distribution), and
U = useful energy (demand level).

The mathematical formulation of MESSAGE constraints ensures that the energy flows are consistent; that is, that (a) not more than the available quantity of resources is consumed, (b) the inflowing amount of energy for each level and each energy carrier is at least equal to the out flowing amount of energy and (c) the demand is met. These three conditions define balance constraints, which are generated by the MESSAGE matrix generator. Their mathematical formulation is documented in section 2.

MESSAGE Inputs

Most MESSAGE inputs can easily be attributed to the three categories of the RES: the primary-energy resources, the conversion technologies and the useful-energy demands. On the resources side, costs, quantities and constraints on the availability of primary-energy sources and resources are required as model inputs. On the demand side, a time series of useful-energy demands must be specified. These either define demands directly or, alternatively, they can be used as the starting point for the iterative MESSAGE-MACRO approach (see section 3), which establishes consistency between energy price and demand changes. The latter is particularly important for the group of mitigation scenarios described in the main part of this book. The biggest set of inputs describes all admissible energy conversion technologies by specifying their cost, performance and availability in time.

Other MESSAGE inputs are not directly related to the cost minimization. Rather, they define the shape of the model, for instance, geographical and temporal subdivisions. The different input categories are presented in more detail in the following subsections.

Time horizon

In its general formulation, MESSAGE is undetermined with regard to the model time horizon and the length of the time steps into which this horizon is divided. The selection of specific values for these parameters depends on the nature of the analytical problem and policy questions to be addressed. Consistent with the focus on long-term sustainability and climate change, the scenarios presented in this book use a time horizon from 1990 to 2100. This time horizon is divided into two initial five-year periods between 1990 and 2000 and into ten-year periods between 2000 and 2100. The selection of 1990 as base year may appear outdated, but it is used nevertheless for reasons of consistency between these and earlier MESSAGE scenarios.

To prevent MESSAGE from 'optimizing' past values, the RES is bounded (that is, forced by constraints) to reflect actual values observed for the years 1990 and 1995. Hence the first point in time for which the optimization is active is the year 2000.[2] The historical build-up of the energy infrastructure prior to 1990 is needed as a model input in order to determine when existing technologies (such as power plants) will go out of operation at the end of their technical and economic lifetime. Further initial inputs are the total available resources in the base year.

All variables of MESSAGE are annual period averages, indexed by the period's initial year.

Eleven world regions

To reflect features of the global energy–economy–environment (E3) system that are specific to world regions and to reflect interregional energy and emission trade, MESSAGE can specify different energy systems for different world regions. Most scenarios presented in this book are based on what can be regarded as a standard disaggregation of the world into 11 regions, as depicted in Figure A.3.[3] In some of the model runs reported in this book, these 11 regions are aggregated into four 'macro' world regions, as follows:

- The OECD90 region groups the countries belonging to the OECD in 1990 (WEU, NAM, PAO).
- The REF ('reforming economies') region aggregates the countries of the Former Soviet Union and Eastern Europe (FSU, EEU).

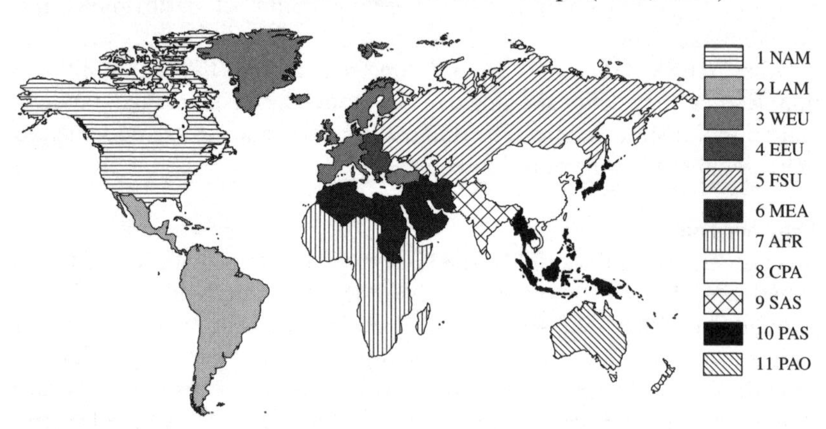

1 NAM North America
2 LAM Latin America and the Caribbean
3 WEU Western Europe
4 EEU Central & Eastern Europe
5 FSU Former Soviet Union
6 MEA Middle East & North Africa
7 AFR Sub-Saharan Africa
8 CPA Centrally planned Asia & China
9 SAS South Asia
10 PAS Other Pacific
11 PAO Pacific OECD

Note: For a list of all countries per world region, see Nakićenović *et al.*, 1998.

Figure A.3 The 11 world regions used in MESSAGE

- The ASIA region represents the developing countries on the Asian continent (SAS, PAS, CPA).
- The ROW region covers the rest of the world, grouping countries in sub-Saharan Africa, Latin America and the Middle East (MEA, AFR, LAM).[4]

The combined OECD90 and REF regions are also referred to as industrialized regions (labelled IND). The ASIA and ROW regions together form the developing regions group (labelled DEV).

Primary-energy extraction

Typical primary-energy resources are hard coal, lignite, crude oil, natural gas and uranium. The model inputs also incorporate, separately for each world region, potentials of the renewable energy sources,[5] wind, geothermal, hydro, solar and biomass. Resource recoverability is given in terms of total resource availability over the time horizon (as constraints on the sum of all annual extraction activities) and in terms of annual extraction (flow) limits of primary energy. Both reflect assumptions on overall economic and technological development, including the cost of exploration and the conversion of resources into reserves.

Energy resources are characterized in MESSAGE by fuel type and energy level in the RES to which the resource is attributed and by grades, that is, categories defined by equal extraction costs. For each grade, the following can be specified:

- total existing resource volume and annual resource extraction in the base year,
- maximum possible rate of growth of resource extraction between two time periods,
- maximum possible resource depletion rate,
- upper limit on the annual extraction.

Real-world examples of grades of crude oil would be offshore and onshore deposits. An example of hard coal would be deposits that can be produced by surface mining and those that require deep mining.

The characteristics most often used to specify assumptions on resources in our MESSAGE scenarios are the energy levels, grades, costs per grade, volume per grade, base year extraction and maximum resource depletion rate. The maximum annual growth of resource extraction is often specified indirectly, by defining a maximum annual growth of the resource extraction technologies (see below).

Energy conversion technologies

Energy technologies are characterized by numerical model inputs describing their economic (for example, cost), technical (for example, conversion efficiencies), ecological (for example, pollutant emissions) and sociopolitical characteristics. An example of the sociopolitical situation in a world region would be the decision by a country or world region to ban certain types of power plants (for example, nuclear plants). Model input data reflecting this situation would be upper bounds of zero for these technologies or, equivalently, their omission from the data set for this region altogether.

MESSAGE provides the option of treating technology data as dynamic quantities. Technology descriptors such as costs, efficiency and maximum utilization per year can therefore be defined as time series, specifying separate values for each time period. This makes it possible to include technological progress in the model without artificially increasing the number of distinct model technologies. Such a simplification can contribute significantly to reducing the model size, which has positive effects on solution times as well as on the interpretability of the results.

Speaking more generally, alternative assumptions on technological change are introduced in MESSAGE by alternative assumptions, (a) for technology cost and performance improvements over time and (b) by alternative assumptions on the first year in which novel technologies become available. Each energy conversion technology is characterized in MESSAGE by the following data:

- Energy inputs and outputs, together with the respective conversion efficiencies. Most energy conversion technologies have one energy input and one output and, thereby, one associated efficiency. But technologies may also use different fuels (either jointly or alternatively) and may have different operation modes and different outputs, which also may have varying shares. An example of different operation modes would be a passout-turbine, which can generate electricity and heat at the same time when operated in cogeneration mode or which can produce electricity only. For each technology, one output and one input are defined as 'main output' and 'main input' respectively. The activity variables of technologies (see pages 191–2) are given in the units of the main input consumed by the technolgy or, if there is no explicit input (as for solar energy conversion technologies), in units of the main output.
- Specific investment costs (for example, per kilowatt, kW) and time of construction as well as distribution of capital costs over construction time.

- Fixed operating and maintenance costs (per unit of capacity, for example, per kW).
- Variable operating costs (per unit of output, for example per kilowatt-hour, kWh, excluding fuel costs).
- Plant availability or maximum utilization time per year. This parameter also reflects maintenance periods and other technological limitations that prevent the continuous operation of the technology.
- Technical lifetime of the conversion technology in years.
- Year of first commercial availability and last year of commercial availability of the technology.
- Consumption or production of certain materials (for example, emissions of kg of CO_2 or SO_2 per produced kWh).
- Limitations on the (annual) activity and on the installed capacity of a technology.
- Constraints on the rate of growth or decrease of annual newly installed capacity and on the growth or decrease of the activity of a technology.
- Technical application constraints, for example, maximum possible shares of wind or solar power in an electricity network without storage capabilities.
- Inventory upon start-up and shutdown: for example, initial nuclear core needed at the start-up of a nuclear power plant.
- Production pattern of the technology in relation to the *load regions* (see pages 180–82).
- Lag time between input and output of the technology.
- Minimum unit size: for example, for nuclear power plants it does not make sense to build plants with a capacity of a few kilowatts power.
- Sociopolitical constraints: for example, ban on nuclear power plants.
- Inconvenience costs. They are specified only for end-use technologies (see the separate description below).

Inconvenience costs of end use technologies

With increasing affluence, consumers of final energy are more likely to demand technologies that are more convenient in their use, even if they cost more than less convenient energy forms. Examples of this empirically observed phenomenon are room heating with gas, electricity or district heat, which are more convenient than heating with coal. The affluent end user does not like to fill up the coal furnace manually and is willing to pay more for a convenient technology. If MESSAGE is to reflect this phenomenon correctly, the model's cost-minimizing behaviour must be modified accordingly. As a model feature to accomplish this task, the concept of inconvenience factors has been introduced in the definition of end-use technologies. The inconvenience factors are specified for each end use technology, time period and world region. The cost entry in the objective function is calculated as the monetary costs, multiplied by the inconvenience factor. The inconvenience factors for a given world region increase with the level of affluence (GDP per capita) in that region, which is a part of a scenario that is calculated by the scenario generator (see section 4.6). Flexible and grid-dependent energy technologies, such as electricity, gas and district heating, have low inconvenience factors.

End-use energy demand

For most scenarios presented in the main part of this book, energy demands were calculated by the Scenario Generator (Gritsevskyi, 1996) and simply transferred into MESSAGE, where they remained unchanged. This was not the case with the mitigation scenarios, for which price-responsive demands were obtained by using MESSAGE-MACRO (see section 3). In those cases, the demands calculated by the Scenario Generator served only as a starting point, which was subsequently modified to reflect energy price increases in the wake of mitigating carbon emissions.[6]

The Scenario Generator calculates the end-use demand under assumptions on the energy efficiency enhancement of end-use technologies. They are given in an abstract way, as an overall energy intensity improvement, expressed in terms of useful energy per GDP. The speed of this improvement is given as part of the so-called 'storyline' of a scenario; that is, it is defined in a way consistent with the assumed economic development. In particular, we assume that higher economic growth favours steeper energy intensity reduction. The results of these assumptions and of the GDP as supplied by the Scenario Generator are MESSAGE inputs consisting of the useful energy trajectories in seven categories (industry electricity, industry other, residential/commercial electricity, residential/commercial other, feedstock, non-commercial, transport).

MESSAGE Outputs and Results

Generally speaking, the output of the MESSAGE model consists of the values of all variables that describe the optimized development of the energy system for all world regions within the given time horizon. In particular, the output describes which mix of technologies and fuels provides the energy supply in each useful-energy demand sector, thus giving a description of the technological development of energy end-use and fuel substitution processes. This picture is made more complete by temporal trajectories for primary, secondary, final and useful energy.

Furthermore, model outputs provide information on the degree to which domestic primary-energy resources are utilized in each time period and overall, as well as on energy imports and exports. Cost information for technologies is given separately for investment, operating and maintenance (O&M), and the price of the input fuels is calculated by the model endogenously. Technology-specific coefficients account for the environmental impact of greenhouse gas (GHG) and pollutant emissions. The newest version of the MESSAGE model includes all six Kyoto-GHGs and pollutant emissions such as NOx and sulphur. In addition, the model permits the analysis of flexible mechanisms such as emissions trading under various equity and burden-sharing assumptions.

The information provided by MESSAGE can be used to facilitate, for example, the following:

- technology selection,
- allocation of resources,
- selection of energy carriers,
- investment decisions,
- optimum plant use,
- analysis of the influence of exogenous (political, ecological) constraints,
- assessment of price trends as a consequence of the system parameters of energy prices, investment costs, environmental requirements and taxation of energy.

2 MATHEMATICAL MODEL DESCRIPTION

This section contains the mathematical formulae defining MESSAGE. All relations that define the structure of a model are given as linear constraints on variables (continuous, integer or binary) and these variables are used to minimize a given objective function, which is the sum of all discounted costs. The possibility of specifying variables as continuous, integer or binary is called the 'mixed-integer option'. In theory, the linearity requirement is a major restriction, but, in practice, nonlinear relationships can be approximated by piecewise linear functions, which are usually sufficient for the given model application.

Owing to the conventional representation of all model relations in matrix form, the variables of such a model are called 'columns', and the constraints, given as equations or inequalities, 'rows'. The constraints and variables are generated from the input data by a part of the MESSAGE programme that is called 'matrix generator'. The variables (columns), which are subject to optimization via the objective function, can be grouped into the following categories:

- Resource extraction variables representing the annual amount of resources extracted in a time period.
- Export and import variables representing the annual export and import of energy carriers.
- Energy conversion activity variables representing the amount of energy converted per year in each time period. Each of these activity variables is associated with a specific energy conversion technology. The units of the activity variables are usually megawatt-years (MWyr) for smaller world regions and gigawatt-years (GWyr) for larger ones. They are also called 'energy flow' variables.
- Capacity variables of energy conversion technologies representing the annual newly installed conversion capacities in a time period (usual unit: MW or GW).
- Stockpile variables representing the quantity of a man-made fuel cumulated over a certain period of time (usual unit: MWyr or GWyr).
- Storage variables representing energy storage input and output, input and output capacity and storage volume.

The model constraints (rows) can be grouped into the following categories:

- Energy flow balances modelling the flow of energy in the energy chain from resource extraction via conversion, transport and distribution up to final utilization.
- Constraints limiting aggregate activities either on an annual or on a cumulative basis, either absolute or in relation to other activities.
- Dynamic constraints setting a relation between the activities of two consecutive time periods.
- Capacity constraints limiting the annual new installation of conversion technology.
- Accounting rows which are not constraining the feasible region of the model application. They therefore do not influence the optimization, and they are only used for calculating selected pieces of information such as pollutant emissions in cases where these are not constrained.

Load Regions

We want to concentrate this description of MESSAGE mainly on those features that were used for the formulation of the scenarios described in this book. Nonetheless, we do not want to exclude important model features from the description. One such feature is the model's ability to include load regions, which we describe in this subsection. Load regions are very useful for detailed local energy system optimizations, but would provide too much detail for global scenarios.

Readers who are only interested in those model capabilities that were actually used for the scenarios in this book can safely skip this subsection and ignore all references to load regions in the description of the model equations further below.

Electricity must be provided by the energy supply system at the very time when it is consumed. Since different amounts of electricity are demanded at different times (of the day and the year), MESSAGE allows the modelling of this situation by dividing each year into an optional number of load regions (time slices), each of them aggregating times with similar demand characteristics. The result of doing so is depicted by a step function as shown in Figure A.4.

For those demands that are divided into load regions, MESSAGE also includes the possibility of energy storage, for example the transfer of energy from night to day or from summer to winter. Energy carriers supplied through a distribution grid are often modelled with a load curve since transport and distribution (and also production in the case of electrical energy) are capital-intensive and since, in general, different types of power plants have different generation costs at different loads. Primary-energy carriers such as coal do not require this precise form of modelling since diurnal or seasonal load fluctuation are of little significance and do not substantially influence capital costs.

Therefore if load regions are defined in a scenario, it is necessary to define for each individual energy carrier whether it is treated as an average quantity or whether the load curve of demand for this energy carrier should be included in modelling. If the main input or output energy carrier of an energy conversion technology is defined to have load regions, then a separate activity variable is generated for each load region by the MESSAGE matrix generator.

End-use technologies are never modelled with load regions. This is because useful-energy demand is not subject to optimization. Any usage patterns are empirical and assumed fixed. MESSAGE therefore represents the end-use technologies by one variable per time period, which produces the required useful energy in the load pattern (of, for example, final-energy supply) needed and requires the inputs in the same pattern. For special technologies such as night storage heating systems, this pattern can be changed to represent the internal storage capability of the system. This representation of end-use technologies has the advantage of reducing the size of the model, because the demand constraints, the activity variables and the

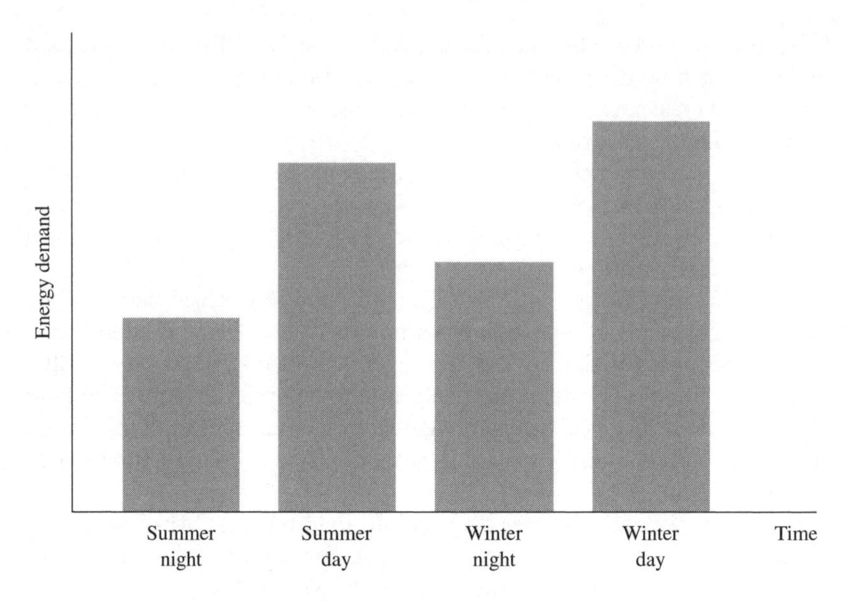

Figure A.4 Example of a semi-ordered load curve

capacity constraints of the end use technologies do not have to be generated for each load region separately.

Demand and Supply Elasticities

In economics, demand and supply are usually considered as nonlinear functions of price, the normal assumption being that demand decreases and supply increases with increasing prices. The concept to quantify these relations is called 'elasticities'. An elasticity expresses the percentage change of one variable as a function of the percentage change of another. For example, a price elasticity of demand of −0.25 means that demand drops by 0.25 per cent when price increases by 1 per cent. MESSAGE can handle demand and supply elasticities by linear approximation.

Notation Conventions used for Documenting the Model Equations

The notation used for the names of variables and constraints in this appendix is the same as in the MPS file of the MESSAGE V model.[7] All names are eight characters long. Admissible characters are alphanumeric characters and dots.

In the following, upper-case alphabetical characters are literal parts of variable and constraint names. They remain fixed for all variable or constraint names in a given class. Lower-case alphabetical characters ('wild cards') hold the place for characters that must be chosen in order to identify a specific variable or constraint. The dots have no specific meaning, they are just used to fill out names to the length of eight characters and they remain a fixed part of all variable or constraint names in the given class. The individual naming rules will be explained when the variable classes are introduced.

In order to keep the notation simple and the mathematical description as short as possible, some descriptors are omitted from the description of the rows in cases where no confusion appears possible; for example, although practically all parameters of MESSAGE are defined as time series (that is, they can assume different values for different time periods), the index of the time period is often omitted in the description below.

In multiregional scenarios such as the ones described in this book, the variable names have an identifier for the world region to which they relate. For example, the resource extraction variables (see below) have an identifier for the world region instead of a dot in the sixth position of the variable name. For simplicity, the mathematical model is presented below only for a model with a single region.

Resource Extraction

Resource extraction variables

The extraction of domestic primary-energy resources (that is, resources available from production within a world region) is modelled by variables that represent the annual quantity extracted in a time period. This variable is linked to the resource extraction technologies as presented on pages 192–7. A division into cost categories (which are called 'grades' in the model) and into supply elasticity classes can be modelled.

The names of resource extraction variables are *Rzrgp..t*, where

- R is the fixed identifier for resource extraction variables,
- z identifies the energy level on which the resource is defined (usually $z = R$, the primary-energy resource level),
- r represents a single character which identifies the resource being extracted (for example, $r = c$, with c being the user-defined identifier for hard coal),
- g represents a single character which identifies the grade (cost category) of resource r,

p is the identifier of the supply elasticity class, which is defined for the resource r and grade g. If no elasticity is defined then p = '.',
t represents a single character that identifies the time period.

To give an illustrative example, *RRcd...f* is the resource extraction variable for coal (identifier c), cost category (grade) d, defined without elasticity classes (.), for the time period f (f represents the sixth time period, that means, 2030–40).[8] The resource variables are energy flow variables and represent the annual extraction of resource r. If several grades are defined, one variable per grade is generated (identifier g in position 4). If supply elasticities are defined for resource extraction, identifier p in position 5 is used.

Resource extraction constraints
In the following subsections the constraints on resource extraction are presented. Constraints are formulated such that all the numerical input parameters are moved to the right-hand side of the constraint. The variables (which are optimized) and their corresponding coefficients are on the left-hand side. Only in some cases, the constraints are presented in a different way if this supports the better understanding of the constraint's meaning.

The availability of a resource can be constrained by the (total and annual) availability per grade as well as by the (total and annual) consumption of all grades of this resource taken together. Additionally, resource depletion and dynamic resource extraction constraints can be modelled. (See the description on pages 197–9.)

Total resource availability per grade The constraints *RRrg....* limit the domestic resource r available from a cost category g (grade) over the whole time horizon. For all combinations of r and g, constraints *RRrg....* are defined as

$$\sum_{p}\sum_{t}\Delta t \times RRrgp..t \leq Rrg - \Delta t_0 R_{rg,0},$$

where the variables subject to optimization are

RRrgp..t = the annual extraction of resource r, cost category g and elasticity class p in time period t,

and the input parameters are

Δt = the length of time period t,

Δt_0 = the number of years between the base year and the first model year,

$R_{rg,0}$ = the extraction of resource r, grade g in the base year, and
Rrg = the total available amount of resource r of grade g in the base year.

Maximum annual resource extraction The constraints $RRr....t$ limit the domestic resources available annually per time period over all cost categories. For each pair of values of r and t, constraint $RRr....t$ is defined as

$$\sum_g \sum_p RRrgp..t \leq Rrt,$$

where the variables subject to optimization are as follows:

$RRrgp..t$ = the annual extraction of resource r, cost category (grade) g and elasticity class p in time period t,

and the input parameters are

Rrt = the maximum amount of domestic resources r, that can be extracted per year in time period t.

Maximum annual resource extraction per grade Constraints $RRrg.a.t$ limit the annual availability of grade g of the domestic resource r. For all combinations of r, g and t, constraints $RRrg.a.t$ are defined as

$$\sum_p RRrgp..t \leq Rrgt,$$

where the variables subject to optimization are as follows:

$RRrgp..t$ = the annual extraction variable of resource r, cost category (grade) g and elasticity class p in time period t,

and the input parameters are

$Rrgt$ = the total amount of resource r, cost category g that is available for extraction in time period t.

Dynamic resource depletion constraints To prevent an unrealistically fast depletion of a resource, the constraints $RRrg.d.t$ can be defined to limit the extraction of a resource r grade g in time period t to a fraction δ^t_{rg} of the total amount still existing in that time period.

For all combinations of r, g and t, constraints $RRrg.d.t$ are defined as

$$\Delta t \sum_p RRrgp..t \leq \delta^t_{rg} \left[Rrg - \Delta t_0 R_{rg,0} - \sum_{\tau=1}^{t-1} \sum_p \Delta\tau \times RRrgp..\tau \right],$$

where the variables subject to optimization are as follows:

$RRrgp..t$ = the annual extraction of resource r, cost category (grade) g and elasticity class p in time period t,

and the input parameters are

δ_{rg}^{t} = the maximum fraction of the resource r, cost category (grade) g, still available at the end of time period $(t-1)$, that can be extracted in time period t,

Rrg = the total amount of resource r, cost category g, that is available, in the base year, for extraction during the model's time horizon,

Δt = the length of time period t,

Δt_0 = the number of years between the base year and the first model year,

$R_{rg,0}$ = the extraction of resource r, grade g in the base year.

Upper dynamic resource extraction constraints The constraints $MRRr...t$ relate the annual extraction of resource r in time period t to the previous time period by specifying a maximum rate of growth and a minimum allowed increment of the extraction activity. For the first time period of the model, the extraction is related to the activity in the base year.

For all combinations of r and t, constraints $MRRr...t$ are defined as

$$\sum_{g,p} RRrgp..t - \gamma_{rt}^{0} \sum_{g,p} RRrgp..(t-1) \leq g_{rt}^{0},$$

where the variables subject to optimization are as follows:

$RRrgp..t$ = the annual extraction of resource r, cost category (grade) g and elasticity class p in time period t,

$RRrgp..(t-1)$ = the same kind of annual extraction for time period $t-1$,

and the input parameters are

γ_{rt}^{0} = the maximum rate of growth of extraction of resource r between time periods $t-1$ and t, and

g_{rt}^{0} = an extra increment, $g_{rt}^{0} > 0$. This allows for the extraction of resource r in time period t also in cases where it was zero in time period $t-1$.

So $g_{rt}^{0} > 0$ ensures that an initial growth is possible and γ_{rt}^{0} ensures that the growth is not limited to an absolute value g_{rt}^{0}, but that a certain increase rate γ_{rt}^{0} is possible in addition.

Lower dynamic resource extraction constraints The constraints $LRRr...t$ relate the annual extraction level of resource r in time period t to the extraction in the previous one by specifying a maximum rate of decrease and a minimum allowed decrement. For the first time period the extraction is related to the activity in the base year.

For all combinations of r and t, constraints $LRRr...t$ are defined as

$$\sum_{g,p} RRrgp..t - \gamma_{rt} \sum_{g,p} RRrgp..(t-1) \geq -g_{rt},$$

where the variables subject to optimization are as follows:

$RRrgp..t$ = the annual extraction variable of resource r, cost category (grade) g and elasticity class p in time period t,

and the input parameters are

γ_{rt} = the maximum rate of decrease of extraction of resource r between time periods $t-1$and t,

g_{rt} = an absolute allowed decrement of extraction of resource r in time period t, $g_{rt} > 0$.

A specification of $g_{rt} > 0$ allows the extraction variables $RRrgp..t$ to decrease at least by the amount g_{rt}. (With $g_{rt} = 0$, only an exponential decay would be admissible, and $RRrgp..t$ could never become zero.)

Dynamic extraction constraints per grade Similar constraints, $MRRrg..t$ and $LRRrg..t$, with analogous parameters can be defined for each grade.

Energy Imports and Exports

Energy imports and exports are modelled by variables that represent, for a world region, the annual quantities of energy imported or exported per year in a time period.

Import and export variables

The import variables are energy flow variables and represent the annual import of the identified energy carrier from another world region. If supply elasticities are defined for the import of this energy carrier and region, one variable per elasticity class (identifier p in position 5) is generated.

For all combinations of z, s, c, p, l and t, the variables $Izscp.lt$ define imports, where

I is a fixed character identifying import variables,

z is a wild-card character for the level on which the imported energy form is defined (usually A for primary energy or X for secondary energy),

s identifies the imported energy carrier,
c is the identifier of the country or region from which the energy carrier 's' is imported,
p is the class of supply elasticity, which is defined for the energy carrier and region, or '.', if no elasticity is defined for this energy carrier and region,
l is the load region identifier if s is modelled with load regions, otherwise '.', and
t identifies the time period.

The export variables are energy flow variables and represent the annual export of the identified energy carrier to another world region. If supply elasticities are defined for the export of this energy carrier and region, one variable per elasticity class (identifier p in position 5) is generated.

For all combinations of z, s, c, p, l and t, the variables $Ezscp.lt$, define exports, where:

E is the identifier for export variables,
z is a wild-card character for the level on which the exported energy form is defined (usually A for primary energy or X for secondary energy),
s identifies the exported energy carrier,
c is the identifier of the country or region to which the energy carrier 's' is exported,
p is the class of supply elasticity, which is defined for the energy carrier and region, or '.', if no elasticity is defined for this energy carrier and region,
l is the load region identifier if s is modelled with load regions, otherwise '.', and
t identifies the time period.

Energy import and export constraints

Maximum total energy imports from a world region The constraints $I.rc....$ limit the imports of fuel r from world region c over the whole time horizon. For all combinations of r and c, constraints $I.rc....$ are defined as

$$\sum_{z,p,t} \Delta t \times Izrcp..t \leq Irc,$$

where the variables subject to optimization are as follows:

$Izrcp..t$ = the annual import variable for imports of fuel r defined on level z from region c, elasticity class p in time period t,

and the input parameters are
Irc = the total import limit for r from region c, and
Δt = the length of time period t in years.

Maximum annual imports into a world region The constraints $I.r....t$ limit the annual imports of a fuel r (from all regions into the given world regions) per time period. For all combinations of r, and t, constraints $I.r....t$ are defined as

$$\sum_{z,c,p} Izrcp..t \leq Irt,$$

where the variables subject to optimization are as follows:

$Izrcp..t$ = the annual import variable for imports of fuel r defined on level z into region c, elasticity class p in time period t,
and the input parameters are
Irt = the annual import limit for r in time period t.

Maximum annual imports from a specific world region The constraints $I.rc...t$ limit the annual imports of energy carrier r from world region c for all elasticity classes p and levels z in time period t. For all combinations of r, c and t, constraints $I.rc...t$ are defined as

$$\sum_{z,p} Izrcp..t \leq Irct,$$

where the variables subject to optimization are as follows:

$Izrcp..t$ = the annual import variable for imports of fuel r defined on level z from region c, elasticity class p in time period t,
and the input parameters are
$Irct$ = the limit on the annual imports from region c, time period t of fuel r.

Dynamic upper constraints on imports The constraints $MI.r...t$ relate, in the same way as for the dynamic resource extraction constraints, the annual import level of fuel r in time period t to the one of the previous time period by a growth parameter and an increment. For all combinations of r and t, constraints $MI.r...t$ are defined as

$$\sum_{z,c,p} Izrcp..t - \gamma^0_{rt} \sum_{z,c,p} Izrcp..(t-1) \leq g^0_{rt},$$

where the variables subject to optimization are as follows:

Izrcp..t = the annual import variable for import of fuel r defined on level z from region c, elasticity class p in time period t,
Izrcp..$(t-1)$ = the same kind of annual import variable for time period $t-1$,
and the input parameters are
γ^0_{rt} = the maximum rate of growth of import of resource r between time periods $t-1$ and t (an input parameter) and
g^0_{rt} = an extra increment to allow for the extraction of resource r in time period t in cases where it was zero in time period $t-1$, $g^0_{rt} > 0$.

Specifying $g^0_{rt} > 0$ ensures that an initial growth is possible, and γ^0_{rt} ensures that an increase rate γ^0_{rt} is permitted in addition.

Lower dynamic import constraints The constraints *LI.r...t* relate the annual import of a fuel in time period t to the annual import in the previous one by a decrease factor and an absolute decrement. For all combinations of r and t, constraints *LI.r...t* are defined as

$$\sum_{z,c,p} Izrcp..t - \gamma_{rt} \sum_{z,c,p} Izrcp..(t-1) \geq -g_{rt},$$

where the variables subject to optimization are as follows:

Izrcp..t = the annual import of resource r from world region c, elasticity class p in time period t,
and the input parameters are
γ_{rt} = the maximum rate of decrease of import of r between time periods $t-1$ and t, and
g_{rt} = the minimum allowed decrement (last size) of import of r in time period t, $g_{rt} > 0$.

A specification of $g_{rt} > 0$ allows the variables *Izrcp..t* to decrease at least by the amount g_{rt}. (With $g_{rt} = 0$, only an exponential decay would be admissible, and *Izrcp..t* could never become zero.)

Lower and upper dynamic import constraints per region The dynamic constraints *MIzrc..t* and *LIzrc..t* similarly relate the annual import from a specific region c of a fuel in the time period t to the annual import in the previous period. Their formulation is completely analogous to those just presented and therefore is omitted.

Constraints on energy exports The exports of fuels can principally be limited in the same way as the imports. In the identifiers of the variables and constraints the '*P*' is just replaced by an '*E*'.

Energy Conversion Technologies

Variables

The energy conversion *activity* variables *zsvd.elt* describe the annual activities of technologies, where

- *z* is the level identifier of the main output of the technology $z \in \{R, A, X, T, F, U\}$. The identifier *U* for *z* denotes the end-use level.[9] This level is handled differently from all other levels: it must be the demand level, and technologies with the main output on this level are defined without load regions;
- *s* represents the identifier for the main energy input (supply) of the technology. If the technology has no explicit input (such as solar technologies), *s* is set to '.';
- *v* is perhaps the most distinctive identifier of the conversion technology[10] (this parameter thus distinguishes technologies with the same input and output, for example, *Xgce....* which in the recent version of MESSAGE is the activity variable of the gas combined-cycle technology, and *Xgfe....* which denotes the gas high-temperature fuel cell activity variable. Both use gas to deliver electricity on level X (secondary energy);
- *d* is the main energy output of the technology (demand sector);
- *e* is the level of reduction of demand due to own-price elasticities of demands (only occurs on the demand level, otherwise (or if this demand has no elasticities) it is set to '.');
- *l* identifies the load region. If the technology is not modelled with load regions, then *l* is set to '.'; and
- *t* identifies the time period.

The activity variable of an energy conversion technology is defined in terms of the annual energy consumption of this technology of the *main input* per time period. The main output and additional outputs and inputs are, as mentioned on pages 176–7, specified in the technology definition sections of the MESSAGE input files. When multiplied with the efficiency ε of the technology, the output of the energy conversion technology can be calculated (output $= \varepsilon \times$ *zsvd.elt*). If a technology has no input, the variable represents the annual production of the main output ($\varepsilon = 1$).

If the level of the main output is *not* equal to 'U' and if at least one of the energy carriers consumed or produced is defined with load regions, the technology is defined with load regions. In this case the activity variables are generated separately for each load region, which is indicated by the additional identifier l in position 7. However, it can be specified that the production of the technology over the load regions is fixed according to a predefined pattern. In this case only one variable is generated for all load regions (for example, the production pattern of solar power plants).

If the model is formulated with demand elasticities and if the activity variables of technologies have a demand as main output that is defined with an elasticity, then separate activity variables are generated for each elasticity class (identifier e in position 6). The energy conversion *capacity* variables $Yzsvd..t$ describe the annual new installation of a technology in a time period, where

Y is the identifier for capacity variables,
z identifies the level on which the main energy output of the technology is defined,
s is the identifier of the main energy input of the technology,
v distinguishes technologies with the same input and output ('the most distinctive identifier of the conversion technology'),
d is the identifier of the main energy output of the technology, and
t is the time period in which the new capacity goes into operation.

The capacity variables represent the annual new installation of a certain capacity of technology v in a time period. The capacity is given in units of the main output of the technology. Technologies can also be modelled without capacity variables. In this case, no capacity constraints and no dynamic constraints on construction are included in the model. Capacity variables of energy conversion technologies can be defined as integer variables, if the solution software includes a mixed-integer option.

Flow balance constraints

There are three main categories of constraints on energy conversion technologies. The *flow balance constraints*, introduced here, ensure that the output of one stage of the energy conversion chain is matched to the demand in the subsequent stage. The energy conversion chain is typically concatenated by (a) the activity variables of the resource extraction $RRrgp..t$, (b) a resource extraction technology $Arvr...t$, (c) a secondary energy production activity $Xrvs..lt$, (d) a transport technology $Tsvs..lt$, (e) a distribution technology $Fsvs..lt$, and (f) an end use energy production activity $Usvd.e.t$. Each such chain potentially contributes to supplying the demand specified in the input of the MESSAGE V model scenario. These constraints are given later in this subsection.

The *dynamic constraints* on energy conversion technology variables limit the activity and/or the capacity of conversion technologies in one time period in relation to the same kind of activities in the time period before. These constraints prevent a model scenario from switching 'instantly' and completely from one technology to another. This would be in implausible contrast to the real world, where technologies are only adopted gradually. Dynamic constraints are given on pages 197–9.

Capacity constraints limit, at each point in time, the activities of each energy conversion technology to the installed conversion capacity of this technology (for example, the electricity generation of power plants must not exceed the installed capacity of this plant). These constraints are defined on pages 199–203. The first flow balance constraint introduced matches resource consumption and use.

Resource consumption balance The $Rr.....t$ constraints link the annual amount of the extracted resource r in a time period to the activities of all technologies v that extract this resource. For all combinations of r and t, constraints $Rr.....t$ are defined as

$$\sum_{g,p} RRrgp..t - \sum_{v} Arvr...t \geq 0,$$

where the variables subject to optimization are as follows:

$RRrgp..t$ is the annual extraction of resource r, cost category (grade) g and elasticity class p in time period t, and

$Arvr...t$ is the energy resource r extraction by technology v in time period t with the energy output on level A.

Primary-energy extraction, export and import balance The equations $Ar.....t$ match the resource extraction plus the import of primary energy to the requirements of central conversion, transport and export. Some technologies, such as nuclear reactors, need inventories of primary energy and also leave a final core that is available at the end of the lifetime.

$$\sum_{v} \varepsilon_{rvr} \times Arvr...t - \sum_{lvs} Xrvs..lt + \sum_{c,p} IArcp..t - \sum_{c,p} EArcp..t +$$

$$\sum_{fvs} \left[\frac{\Delta(t - \tau_{fvs})}{\Delta t} \times \rho(fvs,r) \times YXfvs..(t - \tau_{fvs}) - \frac{\Delta(t+1)}{\Delta t} \times \iota(fvs,r) \times YXfvs..(t+1) \right] \geq 0,$$

where the variables subject to optimization are as follows:

$Arvr...t$ = the activity of technology v extracting resource r; in other words, this is the annual demand for domestic primary energy r in time period t,

$Xrvs..lt$ = the activity variable of the central conversion technology v, converting the primary energy r to the secondary-energy form s in load region l and time period t, with the energy output defined on level X,

$IArcp..t$ = the import variables for import on the primary-energy level A,

$EArep..t$ = the export variables for exports on the primary-energy level A,

$YXfvs..t$ = the annual newly installed technology v with outputs on level X in time period t,

$YXfvs..(t-\tau)$ = the annual new installation of technology v in time period $(t-\tau)$,

and the input parameters are

ε_{rvr} = the efficiency of technology v in extracting resource r (this is usually 1.); this parameter can be used to describe extraction losses;

τ_{fvs} = the plant life of technology v in time periods (depending on the lengths of the time periods covered);

$\iota(fvs,r)$ = the amount of fuel r that is needed when technology v goes into operation (usually this is the first core of a nuclear reactor). It has to be available in the time period before technology v goes into operation; the normal unit is kWyr/kW;

$\rho(fvs,r)$ = the amount of fuel r that becomes available after technology v goes out of operation (for a nuclear reactor this is the last core that goes to reprocessing). The unit is the same as $\iota(fvs,r)$,

Δt = the length of time period t in years;

$\Delta(t-\tau)$ = the length of time period $(t-\tau)$ in years.

Central-conversion balance The equations $Xs....lt$ describe the balances of secondary energy in time period t. In principle, the central-conversion balance is formulated in the same way as the two previous ones. It matches the secondary-energy supply; that is, the production of central conversion technologies to the demands for secondary energy, which is calculated endogenously by the model and formulated as the requirements by the transmission systems. Secondary-energy imports and exports of secondary energy are usually assigned to level X and are therefore included here. For all combinations of s, l and t, constraints $Xs....lt$ are defined as

$$\sum_{rv} \varepsilon_{rvs} \times Xrvs..lt + \sum_{rv\sigma} \beta^{s}_{rv\sigma} \times Xrv\sigma..lt - \sum_{v} Tsvs..lt +$$
$$\sum_{c,p} IXscp.lt - \sum_{c,p} EXscp.lt \geq 0,$$

where the variables subject to optimization are as follows:

$Xrvs..lt$ = the activity variables of central-conversion technology v in load region l and time period t with output on level X. If the secondary-energy form s is not defined with load regions (that is, l = '.') but the activity of technology v exists for each load region, this equation will contain the sum of the activity variables of technology v over the load regions,

$Tsvs..lt$ = the activity variable of the transportation technology v, with input on level X and

$IXscp.lt$, $EXscp.lt$ = the variables describing, respectively, the import and export of secondary energy,

and the input parameters are

ε_{rvs} = the efficiency of technology v in converting energy carrier r into secondary energy form s,

$\beta^{s}_{rv\sigma}$ = the efficiency of technology v in converting energy carrier r into the byproduct s of technology v.

Transmission or transportation balance The constraints $Ts....lt$ match supply and demand for final energy. These constraints are the simplest of all energy balance constraints of MESSAGE. They match the output of transmission technologies to the (endogenously calculated) final-energy requirements of distribution systems. The difference between the formulation of these constraints and those formulated for the other levels (F, X and A) is not by deliberate design, but emerges from the simplicity of energy transportation (that is, transportation technologies usually do not have byproducts).[11] As for level F, the constraints are defined for all load regions whenever the load regions are defined for the fuel. For all combinations of s, l and t, constraints $Ts....lt$, are defined as

$$\sum_{v} \varepsilon_{svs} \times Tsvs..lt - \sum_{v} Fsvs..lt \geq 0,$$

where the variables subject to optimization are as follows:

$Tsvs..lt$ = the activity of the transportation technology v,

$Fsvs..lt$ = the activity of the distribution technology in load region l and time period t,

and the input parameters are

ε_{svs} = the efficiency of technology v in distributing s.

Distribution balance The constraints $Fs....lt$ match the supply of final energy, given as the deliveries of the distribution technologies with the final-energy requirements of the end-use technologies. These constraints are generated for each load region if the final-energy form is modelled with load regions. For all combinations of s, l and t, constraints $Fs....lt$ are defined as

$$\sum_{v} \varepsilon_{svs} \times Fsvs..lt - \sum_{vd} \eta_{s,v,d,l,t} \times \sum_{e=0}^{e_d} Usvd.e.t -$$

$$\sum_{\sigma vd} \overline{\beta}^{s}_{\sigma vd} \times \eta_{\sigma,v,d,l,t} \times \sum_{e=0}^{e_d} U\sigma vd.e.t \geq 0,$$

where the variables subject to optimization are as follows:

$Fsvs..lt$ = the activity variable of the distribution technology in load region l and time period t,

$Usvd.e.t$ = the activity variable of end use technology v in time period t and elasticity class e,

and the input parameters are

ε_{svs} = the efficiency of technology v in distributing s,

$\overline{\beta}^{s}_{\sigma vd}$ = the use of fuel s relative to fuel σ (the main input) by technology v, and

$\eta_{s,v,d,l,t}$ = the fraction of demand for d occurring in load region l.[12]

As all the activities are defined in units of the energy inputs into the final-conversion technologies, the efficiency factors $\varepsilon_{s,v,s}$ must be used to calculate the output of the distribution technology $Fsvs..lt$.

End-use demand balance At long last, the constraints $Ud.....t$ make sure that the given end-use demand is satisfied by the supply provided by the end-use technologies. In principle, each of the predefined levels can be chosen as demand level, but here, as in all MESSAGE scenarios described in this book, end-use demand is defined in terms of useful energy. As explained on pages 180–82, end use level 'U' is not modelled with load regions. For all combinations of d and t, constraints $Ud.....t$ are defined as

$$\sum_{sv} \varepsilon_{svd} \times \sum_{e=0}^{e_d} k_e \times Usvd.e.t + \sum_{sv\delta} \beta^{d}_{sv\delta} \times \sum_{e=0}^{e_\delta} k_e \times Usv\delta.e.t \geq Udt,$$

where the variables subject to optimization are as follows:

Usvd.e.t = the activity variable for the production of end-use energy form d, elasticity class e from input s by technology v in time period t,

and the input parameters are

Udt = the annual demand for d in time period t,

ε_{svd} = the efficiency of end-use technology v in converting final-energy form s to the end-use energy form d,

$\beta^{d}_{sv\delta}$ = the efficiency of end-use technology v in producing byproduct d from s (δ is the main output of the technology),

e_d = the number of steps of demand reduction modelled for own-price elasticities of demand d, and

k_e = the factor giving the relation of initial demand *Usvd.0.t* for end-use energy form d to the demand reduced to a lower level *Usvd.e.t* due to the demand elasticity ($k_e = Usvd.0.t \,/\, Usvd.e.t$).

The activity *Usvd.e.t* is given in units of the energy input s. Therefore the efficiency factor ε_{svd} must be used to calculate the output of end-use energy form d produced by the activity *Usvd.e.t*.

Dynamic constraints

Expressed in general terms, dynamic constraints are used to reflect the inertia of real-world energy systems in the model. Dynamic constraints ensure that a technology is only gradually introduced or removed from the energy system. As technologies can be modelled with or without capacity variables, dynamic constraints can be defined for activity variables as well as for the variables describing the installation of new capacity.

Upper dynamic constraints on activity variables The constraints *Mzsvd..t* relate the production of a technology in one time period to the production in the previous one. If the technology in question is defined with load regions, the sum over the load regions is included in the constraint. For all combinations of z, s, v, d and t, constraints *Mzsvd..t* are defined as

$$\sum_{l} \varepsilon_{svd} \times \left[zsvd..lt - \gamma a^{0}_{svdt} \times zsvd..l(t-1) \right] \leq ga^{0}_{svdt},$$

where the variables subject to optimization are as follows:

zsvd..lt = the activity of technology v in load region l,

and the input parameters are

γa^0_{svdt} = the maximum growth rate,
ga^0_{svdt} = the increment (the increment must be given in units of main output), $ga^0_{svdt} > 0$.

Lower dynamic constraints on activity variables Whereas the previous group of constraints prevent an overly fast introduction of technologies, the constraints *Lzsvd..t* prevent a sudden drop-out of technologies. For all combinations of *z*, *s*, *v*, *d* and *t*, constraints *Mzsvd..t* are defined as

$$\sum_l \varepsilon_{svd} \times [zsvd..lt - \gamma a_{svdt} \times zsvd..l(t-1)] \geq -ga_{svdt},$$

where the variables subject to optimization are as follows:

zsvd..lt = the activity of technology *v* in load region *l*,
and the input parameters are
γa_{svdt} = the maximum decrease rate
ga^0_{svdt} = the decrement, $ga_{svdt} > 0$.

Upper dynamic constraints on construction variables In complete analogy to the corresponding dynamic constraints on activity variables, constraints *MYzsvd.t* relate the amount of annual new installations of a technology in a time period to the annual construction during the previous time period. For all combinations of *z*, *s*, *v*, *d* and *t*, constraints *MYzsvd.t* are defined as

$$Yzsvd..t - \gamma y^0_{svd,t} \times Yzsvd..(t-1) \leq gy^0_{svd,t},$$

where the variables subject to optimization are as follows:

Yzsvd..t = the annual new installation of technology *v* in time period *t*,
and the input parameters are
γy^0_{svdt} = the maximum growth rate of the annual installations of technology *v* per time period,
gy^0_{svdt} = the minimum allowed increment (initial size) for the installations of the new technologies *v*, $gy^0_{svdt} > 0$.

The upper dynamic constraints have become commonly known as market-penetration constraints.

Lower dynamic constraints on construction variables Analogous to both market-penetration constraints just defined and the lower dynamic constraints on activity variables, the constraints *Lyzsvd.t* prevent a sudden discontinuation of installing new capacities of technologies. For all combinations of z, s, v, d and t, constraints *Lyzsvd.t* are defined as

$$Yzsvd..t - \gamma y_{svd,t} \times Yzsvd..(t-1) \leq -gy_{svd,t},$$

where the variables subject to optimization are as follows:

$Yzsvd..t$ = the annual new installation of technology v in time period t, and the input parameters are

$\gamma y_{svd,t}$ = the maximum rate of decrease for the annual installation per time period for the construction of technology v, and

$gy_{svd,t}$ = the minimum allowed decrement (last step) allowing technologies to go out of the market.

Capacity constraints

Capacity constraints *Czsvd.lt* on energy conversion technologies limit the utilization of a technology in relation to the capacity actually installed. These capacity constraints are generated for all conversion technologies modelled with capacity variables.

Technologies without load regions For technologies without load regions (that is, technologies for which neither input nor output is modelled with load regions) the annual production *zsvd...t* is related to the total installed capacity simply by the plant factor. For these technologies, the plant factor must be given as the fraction of the year during which the technology actually operates. All end-use technologies (technologies with main output level '*U*') are modelled in this way. If the technology is an end-use technology, the sum over the elasticity classes will be included in the capacity constraint. In this case, for all combinations of z, s, v, d and t, constraints *Czsvd..t* are defined as

$$\varepsilon_{svd} \times zsvd\ldots t - \sum_{\tau=t-\tau_{svd}}^{\min(t,\kappa_{svd})} \Delta(\tau) \times \pi_{svd} \times f_i \times Yzsvd..\tau \leq hc^t_{svd} \times \pi_{svd},$$

where the variables subject to optimization are as follows:

$zsvd\ldots t$ = the activity of conversion technology v with input energy form s and output energy form d defined on level z in time period t,

$Yzsvd..t$ = the annual new installation of the conversion technology v with input energy form s and output energy form d defined on level z in time period t

and the input parameters are

ε_{svd} = the efficiency of technology v in converting the main energy input, s, into the main energy output, d,

κ_{svd} = the last time period in which technology v can be constructed,

π_{svd} = the plant factor of technology v, meaning the fraction of time (a year) in which the technology actually operates,

$\Delta\tau$ = the length of time period τ in years,

τ_{svd} = the plant life of technology v, expressed in time periods,

hc^t_{svd} = represents the installations built before the base year under consideration that are still in operation in the first year of time period t, and

f_i = is equal to 1 if the capacity variable is continuous, and represents the minimum installed capacity per year (unit size) if the variable is an integer.

Technologies with load regions and free production pattern[13] If a technology has at least one input or output with load regions, the activity variables and capacity constraints are generated separately for each load region. This is to say that, for each load region, the activity of a technology can be optimized separately, and its activity in each load region is constrained by the installed capacity. The plant factor of such a technology must be given as the fraction of time during which the technology can operate in peak operation mode (in general, this is the availability factor). For all combinations of z, s, v, d, l and t, constraints $Czsvd.lt$ are defined as

$$\frac{\varepsilon_{svd}}{\lambda_l} \times zsvd..lt - \sum_{\tau=t-\tau_{svd}}^{\min(t,\kappa_{svd})} \Delta(\tau) \times \pi_{svd} \times f_i \times Yzsvd..\tau \leq hc^t_{svd} \times \pi_{svd},$$

where the variables subject to optimization are as follows:

$zsvd..lt$ = the activity of conversion technology v with input energy form s and output energy form d defined on level z in time period t,

$Yzsvd..t$ = the annual new installation of the conversion technology v with input energy form s and output energy form d defined on level z in time period t

and the input parameters are

λ_l = the length of load region l as fraction of the year,

ε_{svd} = the efficiency of technology v in converting the main energy input, s, into the main energy output, d,

κ_{svd} = the last time period in which technology v can be constructed,

π_{svd} = the plant factor of technology v, given as the fraction of time during which the technology can operate in peak operation mode,

$\Delta\tau$ = the length of time period τ in years,
τ_{svd} = the plant life of technology v, expressed in time periods,
hc^t_{svd} = the installations built before the base year under consideration that are still in operation in the first year of time period t, and
f_i = 1 if the capacity variable is continuous, and represents the minimum installed capacity per year (unit size) if the variable is an integer.

Technologies with load regions and a fixed production pattern If a technology has at least one input or output with load regions, but where the production pattern over the load regions is predefined, the capacity constraint is generated for only the load region where the production pattern assigns the highest production activity. The plant factor must be given for the load region with the highest assigned production activity. For all combinations of z, s, v and t, constraints $Czsvd..t$ are defined as

$$\frac{\varepsilon_{svd} \times \pi(l_m,svd)}{\lambda_{lm}} \times zsvd...t - \sum_{\tau=t-\tau_{svd}}^{\min(t,\kappa_{svd})} \Delta(\tau) \times \pi_{svd} \times f_i \times Yzsvd..\tau \leq hc^t_{svd} \times \pi_{svd},$$

where the variables subject to optimization are as follows:

$zsvd...t$ = the activity of conversion technology v with input energy form s and output energy form d defined on level z in time period t,
$Yzsvd..t$ = the annual new installation of the conversion technology v with input energy form s and output energy form d defined on level z in time period t

and the input parameters are

l_m = the load region with the maximum capacity use if the production pattern over the year is fixed,
λ_{lm} = the length of load region l_m,
$\pi(l_m,svd)$ = the share of output in the load region l_m,
ε_{svd} = the efficiency of technology v in converting the main energy input, s, into the main energy output, d,
κ_{svd} = the last time period in which technology v can be constructed,
$\Delta\tau$ = the length of time period τ in years,
τ_{svd} = the plant life of technology v, expressed in time periods,
hc^t_{svd} = the installations built before the base year under consideration that are still in operation in the first year of time period t, and
f_i = 1 if the capacity variable is continuous, and represents the minimum installed capacity per year (unit size) if the variable is an integer.

Technologies with varying inputs and outputs The activity variables of different technologies can be linked to the same capacity variable, which allows leaving the choice of the activity variable used with a given capacity subject to optimization. For example, if a pass-out turbine generates electricity and heat in cogeneration in the main operation mode (*zsvd...t*), and electricity only in the alternative mode (*zsv'd...t*), then both activities are linked to the same installed capacity (the existing power plant). The installed capacity *Yzsvd...t* is given, as stated above (see page 192), in units of the main output in the main operation mode ($\varepsilon_{svd} \times zsvd...t$), here electricity, whereas the activity variables (*zsvd...t*, *zsv'd...t*) are given in units of the main input, here gas. In order to link the activity variables (*zsvd...t*, *zsv'd...t*) to the same capacity variable *Yzsvd...t*, the factor $rel_{sv'd}^{svd}$ must be defined as given below.

The following constraints are described only for the case of technologies without load regions. The constraints with load regions are constructed in complete analogy. The constraints link the different activity variables to one capacity variable. For all combinations of *z*, *s*, *v*, *d* and *t*, constraints *Czsvd...t* are defined as

$$\sum_{\sigma v'\delta} rel_{\sigma v'\delta}^{svd} \times \varepsilon_{\sigma v'\delta} \times z\sigma v'\delta\ldots t - \sum_{\tau=t-\tau_{svd}}^{\min(t,\kappa_{svd})} \Delta(\tau) \times \pi_{svd} \times f_i \times Yzsvd\ldots\tau \leq hc_{svd}^{t} \times \pi_{svd},$$

where the variables subject to optimization are as follows:

zsvd...t = the activity of conversion technology *v* with input energy form *s* and output energy form *d* defined on level *z* in time period *t*,
Yzsvd..t = the annual new installation of the conversion technology *v* with input energy form *s* and output energy form *d* defined on level *z* in time period *t*,

and the input parameters are

$rel_{\sigma v'\delta}^{svd}$ = the capacity of the main output of the main operation mode *svd* relative to the capacity of output of the alternative operation mode $\sigma v'\delta$ for technology *v*,
ε_{svd} = the efficiency of technology *v* in converting the main energy input, *s*, into the main energy output, *d*,
κ_{svd} = the last time period in which technology *v* can be constructed,
$\Delta\tau$ = the length of time period τ in years,
π_{svd} = the plant factor of technology *v*, meaning the fraction of time (a year) in which the technology actually operates,

f_i = 1 if the capacity variable is continuous, and represents the minimum installed capacity per year (unit size) if the variable is integer, and

hc^t_{svd} = the installations built before the base year under consideration that are still in operation in the first year of time period t.

Stockpiles

Man-made fuels such as reprocessed nuclear fuel accumulate over time and stockpiles are generated. These are described by the variables $Qfb....t$.

Variables

For all fuels f that are defined as generating stockpiles and all time periods t, the variables $Qfb....t$ are generated, where

Q is the identifier of the stockpile variables,
f identifies the fuel with stockpile,
b distinguishes the variable from the equation, and
t is the time period identifier.

The stockpile variables represent the amount of fuel f that is transferred from time period t into time period $t+1$. Note that these variables do not represent annual flows, but average stocks in a time period as a whole. Stockpile constraints are defined as a separate level, Q. For all other energy carriers, any overproduction that may occur in a time period is lost.

Constraints (equations)

For all fuels f for which stockpiles are to be included in MESSAGE, constraints $Qf.....t$ are defined as

$$Qfb....t - Qfb....(t-1) = -\sum_{z,v,d,l} \Delta t \times zfvd..lt - \sum_{z,v,\phi,d,l} \bar{\beta}^f_{\phi vd} \times z\phi vd..lt +$$
$$\sum_{z,s,v,l} \varepsilon_{svf} \times zsvf..lt + \sum_{z,s,v,\phi,l} \beta^f_{sv\phi} \times zsv\phi..lt +$$
$$\sum_{z,s,v,d} [\Delta(t-\tau_{svd}) \times \rho(svd,f) \times Yzsvd..(t-\tau_{svd}) -$$
$$\Delta t \times \iota(svd,f) \times Yzsvd..(t)],$$

where the variables subject to optimization are as follows:

$Qf.....t$ = the stockpile variable of fuel f in time period t defined on the stockpile level Q,

$zfvd..lt$ = the annual input of fuel f for technology v in load region l and time period t (l is '.' if v does not have load regions),
$Yzfvd..t$ = the annual new installation of technology v in time period t
and the input parameters are
f = the identifier of the man-made fuel (for example plutonium, U_{233})
τ_{svd} = the plant life of technology v in time periods,
$\iota(svd,f)$ = the 'first inventory' of technology v of f (relative to capacity of main output),
$\rho(svd,f)$ = the 'last core' of f in technology v,
Δt = the length of time period t in years,
$\overline{\beta}^{f}_{\phi vd}$ = the use of fuel f relative to fuel ϕ (the main input) by technology v,
$\beta^{f}_{sv\phi}$ = the efficiency of end-use technology v in producing byproduct f from s (ϕ is the main output of the technology),
ε_{svd} = the efficiency of technology v in converting the main energy input, s, into the main energy output, d.

User-defined Relations

One of the most powerful modules of MESSAGE V is one that provides the possibility for the user to define freely constraints involving all types of technology-related variables. The user can, among other things, limit one technology in relation to some other technologies (for example, a maximum share of wind energy that can be handled in an electricity network), give exogenous limits on sets of technologies (for example, a common limit on all technologies emitting SO_2) or define additional constraints between the production and the installed capacity of a technology (for example, to ensure take-or-pay clauses in international gas contracts, which oblige customers to consume a minimum share of the contracted level during summer months).

A rather simple example of a user-defined relation is the constraint USO2, which limits annual SO_2 emissions for scenarios that neither use load regions nor demand elasticities. The constraint USO2 is defined as

$$\sum_{zrvs} ro_{rvs} \times zrvs\ldots t \quad \begin{cases} free \\ \leq rhs^{t} \end{cases}$$

where the factors ro_{rvs}, represent SO_2 emissions in tons per unit of energy input of the energy conversion technology *rvs*. They must be specified in the definition of each energy conversion technology. In the definitions of the USO2 user-defined relation, it would be defined whether there is an upper boundary rhs^t on the allowed SO_2 emissions or whether there is no boundary ('free' constraint). In either case, an entry quantifying specific costs (costs per ton of SO_2 emissions) can be specified for the objective function. Depending on these costs, the optimization will put more or less weight on SO_2 emissions.

Generally speaking, for each technology, the user can specify coefficients for any number of user-defined relations. The coefficients may relate to the activity variables, the annual new installation of capacity or the total installed capacity. The user-defined relations just add up the activities (after multiplying them with the given coefficients) of all variables that are defined to be represented in this constraint.

A user-defined relation can apply to inputs and/or outputs of energy conversion variables, it can apply only for a certain load region or for all load regions, it can have lower and/or upper bounds, it may be defined as an equation or as unlimited ('free'), and it may generate cost entries in the objective function. It is possible to define relations between variables of two consecutive time periods and relations with special handling of demand elasticities, which allows including the effects of investments in energy-saving technology.

Objective Function and Cost Counters

Within the feasible solution space, which is defined by the Reference Energy System (RES), MESSAGE selects a solution by optimizing the objective function, which represents the decision criteria and preferences for the selection of energy carriers and technologies. In any conventional techno-economic energy model, the objective function is total discounted costs, which are to be minimized. This is also the default in MESSAGE, which minimizes total energy system costs, discounted over the entire time horizon of the given scenario.[14]

Cost accounting rows

Different types of costs (that is, entries for the objective function) can be accounted for separately in built-in accounting rows. These rows can be generated per time period and/or for the whole time horizon and contain the sum of the undiscounted costs. They can also be limited. The implemented cost types are as follows:

CCUR fixed (related to the installed capacity) and variable (related to the production) operation and maintenance costs,

CCAP investment costs, given as annualized costs distributed over a number of time periods according to the lifetime of the corresponding technologies (*CCAP* shows the share of investments in the respective time period),

CRES domestic-fuel consumptions,
CAR costs related to the user-defined relations, e.g., USO2 (see the description above),
CRED costs of reducing demands due to demand elasticities, only related to technologies supplying the demands directly,
CIMP import costs,
CEXP revenues from exports, and
CINV total investments given for each time period (not annualized).

The objective function

In its standard form, the objective function *FUNC* contains the sum of all discounted costs, that is, all kinds of costs that are accounted for in MESSAGE. Discounting of all costs related to operation (that is, resource use, operation costs, costs of demand elasticities and so on) is from the *middle* of the current time period to the first year. Costs related to construction are discounted from the *beginning* of the current time period to the first year. MESSAGE also foresees the possibility of distributing total investments over more than one time period before or within the current one. This distribution can also be defined for user-defined relations.

The objective function *FUNC* has the following general form:

$$\sum_t \Bigg[\beta_m^t \Delta t \Bigg\{ \sum_{svdl} zsvd\ldots lt \times \varepsilon_{svd} \times \Bigg[ccur(svd,t) + \sum_m ro_{svd}^{mlt} \times car(ml,t) \Bigg] +$$

$$\sum_{svd} \varepsilon_{svd} \times \sum_{e=0}^{e_d} Usvd.e.t \times \varepsilon_{svd} \times \Bigg[\kappa_e \times (ccur(svd,t) + \sum_m ro_{svd}^{mt} \times$$

$$car(m,t)) + cred(d,e) \Bigg] + \sum_{svd} \sum_{\tau = t - \tau_{svd}}^{t} \Delta\tau \times Yzsvd\ldots\tau \times$$

$$cfix(svd,\tau) + \sum_r \Bigg[\sum_{glp} Rzrgp.lt \times cres(rgpl,t) +$$

$$\sum_{clp} (Izrcp.lt \times cimp(rcpl,t) - Ezrcp.lt \times cexp(rcpl,t)) \Bigg] \Bigg\} +$$

$$\beta_b^t \times \Bigg\{ \sum_{svd} \sum_{r=t}^{t+t_d} \Delta(t-1) \times Yzsvd\ldots\tau \times \Bigg[ccap(svd,\tau) \times fri_{svd}^{t_d - \tau} +$$

$$\sum_m rc_{svd}^{mt} \times car(m,t) \times fra_{svd,m}^{t_d - \tau} \Bigg] \Bigg\} \Bigg],$$

where the variables subject to optimization are as follows:

Usvd.e.t = the annual consumption of fuel s of end-use technology v in time period t and elasticity class e,
Yzsvd..t = the annual newly built capacity of technology v in time period t,
Rzrgp.lt = the annual consumption of resource r, grade g, elasticity class p in load region l and time period t,
Izrcp.lt = the annual import of fuel r from region c in load region l, time period t and elasticity class p (if r has no load regions, then l = '.'),
Ezrcp.lt = the annual export of fuel r to region c in load region l, time period t and elasticity class p (if r has no load regions, then l = '.'),

and the input parameters are

Δt = the length of time period t in years,

$$\beta_b^t = \prod_{i=1}^{t-1} \left[\frac{1}{1 + \frac{dr(i)}{100}} \right]^{\Delta i}$$

the discount factor for *investment costs*. All entries in the objective function are discounted from the beginning of that period to the first year, if they represent investment costs into new capacity (capacity variables). Note that the symbol Π denotes the product of the terms with index i,

$$\beta_m^t = \beta_b^t \left[\frac{1}{1 + \frac{dr(i)}{100}} \right]^{\frac{\Delta t}{2}}$$

the discount factor for entries in the objective function related to *activity variables*. They are discounted from the middle of the respective period to the first year,

$dr(i)$ = the discount rate in time period i in per cent,
zsvd..lt = the annual consumption of technology v of fuel s in load region l and time period t (if v has no load regions, l = '.'),
ε_{svd} = the efficiency of technology v in converting fuel s to energy form d,
$ccur(svd,t)$ = the variable operation and maintenance costs of technology v (per unit of main output) in time period t,
ro_{svd}^{mlt} = the relative factor per unit of utilized capacity of technology v for relational constraint m, in load region l, time period t,

$car(m,t)$ = the coefficient for the objective function related to the user-defined relation m in time period t,
$car(ml,t)$ = the same for load region l, if relation m has load regions,
κ_e = the factor giving the relation of total demand for d to the demand reduced to level e, due to the elasticity,
$cred(d,e)$ = the cost associated with reducing the demand for d to elasticity level e,
$cfix(svd,t)$ = the fixed operation and maintenance cost of technology v that was built in time period t (given per unit of installed capacity),
$ccap(svd,t)$ = the specific investment cost of technology v in time period t (given per unit of main output),
frl^n_{svd} = the share of this investment to be paid over n time periods before the first year of operation,
rc^{mt}_{svd} = the relative factor per unit of newly built capacity of technology v for user-defined relation m in time period t,
$fra^n_{svd,m}$ = the share of the relative amount of the user-defined relation m that occurs over n time periods before the first year of operation (this can, for example, be used to account for the use of steel in the construction of solar towers over the time of construction),
$cres(rgpi,t)$ = the cost of extracting resource r, grade g, elasticity class p in time period t and load region l (this should only be given if the extraction is not modelled explicitly),
$cimp(rcpl,t)$ = the cost of importing energy form r in time period t from region c in load region l and elasticity class p,
$cexp(rcpl,t)$ = the revenue from exporting r in time period t to region c in load region l and elasticity class p.

Advanced Features

In combination with an appropriate solver such as MINOS (Murtagh and Saunders, 1983), MESSAGE can be applied to problems with a partly nonlinear objective function or with nonlinear constraints. The requirements are that the functions be differentiable and convex with respect to the solution space. In order to use a nonlinear objective or nonlinear constraints the user must identify the variables that are to be included with nonlinear coefficients in the input file and to supply MINOS with the nonlinear part of the constraints or objective function and the first derivatives.

MESSAGE also supports two types of multi-objective optimization. First, weights can be put on all entries of user-defined relations and technology variables in the objective function. The results of optimization of the objective function will reflect these weights. This is the standard multi-objective approach. The weights must be comparable to each other and are chosen by the user.

The second, reference-trajectory optimization, approach (Grauer *et al.*, 1982) allows defining 'reference levels' for each objective function in the model. The advantage of this approach is that no weights need to be defined, and that the reference levels are given in the natural units of the objective, for example in tons of SO_2 emitted. The optimization attempts to reach the reference levels of all objectives. If this turns out infeasible, the model will attempt to reach these levels as closely as possible with a Pareto-optimal solution. If the reference levels can be reached, the model will again find a Pareto-optimal solution, this time one that will either reach these levels or 'outperform' them. In either of these two cases, the algorithm uses its own metric to measure 'closeness'.[15] The normal way of using MESSAGE in this mode therefore consists of performing a series of runs between which the reference levels can be redefined in the light of earlier solutions offered by the algorithm.

3 MESSAGE-MACRO

MESSAGE-MACRO links MESSAGE with a non-linear macroeconomic model (MACRO). MACRO has its roots in a long series of models by Manne and others. The latest model in this series is MERGE 5 (Manne and Richels, 2003). MACRO is a macroeconomic model maximizing the intertemporal utility function of a single representative producer–consumer in each world region. The optimization result is a sequence of optimal savings, investment and consumption decisions. The main variables of the model are the capital stock, available labour, and energy inputs, which together determine the total output of an economy according to a nested CES (constant elasticity of substitution) production function. Energy demand in two categories (electricity and non-electric energy) is determined within the model, and is consistent with energy supply curves, which are inputs to the model (from MESSAGE). To accommodate the two energy sectors of MACRO, the seven demand sectors of MESSAGE were aggregated into the two that are required.

The model's most important driving input variables are the projected growth rates of total labour, that is, the combined effect of labour force and labour productivity growth and the annual rates of reference energy intensity reduction. Labour growth is also referred to as reference GDP growth. In the absence of price changes, energy demands grow at rates that are the approximate result of GDP growth rates, reduced by the rates of overall energy intensity reduction. Price changes can alter this path significantly.

The vehicle transporting the information between the two submodels is supply curves derived from MESSAGE. These supply curves are input to MACRO, which returns a new set of demands, consistent with MACRO's production function, to MESSAGE. These steps are repeated until convergence is achieved. That is to say, the iteration stops as soon as the new set of demands as produced by MACRO are close to those provided as inputs by MESSAGE.

For a more comprehensive description of MESSAGE-MACRO see Messner and Schrattenholzer (2000).

4 THE SCENARIO GENERATOR (SG)

The Scenario Generator (Gritsevskyi, 1996) is a simulation model to help formulate scenarios of economic and energy demand development for 11 world regions analysed by MESSAGE. Its main objective is to allow the scenario formulation and documentation of key scenario assumptions, and to provide common, consistent input data for MESSAGE and MACRO.

Within the Scenario Generator (SG) there are, first, consistent sets of economic and energy data for the base years 1990 and 2000, plus time series of such data for prior years. Second, the SG contains a set of regression equations estimated using the economic and energy data sets. These equations represent key relationships between economic and energy development, based on empirical data, that can be used selectively in formulating particular scenarios. To allow adjustments for different storylines and variants, all important variables are formulated so that a user can overwrite the values suggested by the equations of the SG.

Inputs to the SG are future population trajectories for 11 world regions used by MESSAGE plus key parameters determining regional per capita GDP growth. The SG first calculates growth rates of total GDP for each world region. Second, it calculates total final-energy trajectories for each region by combining the population and per capita GDP growth trajectories with final-energy intensity profiles based on the SG's set of empirically derived equations. The resulting final-energy demands are then disaggregated, again based on combining regional per capita income growth with the SG's set of empirically derived equations, into the seven demand sectors used by MESSAGE and listed below. In the list, 'specific' energy demands are those that require electricity (or its substitutes such as, in the long term, hydrogen). 'Non-specific' energy demands are mainly thermal requirements that can be fulfilled by any energy form. The seven sectors are industrial specific, industrial non-specific, residential/commercial specific, residential/commercial non-specific, transport, feedstocks, and non-commercial (for example, fuelwood).

5 A BRIEF SUMMARY OF MESSAGE'S APPLICATIONS

Global Energy Perspectives, 1998

In 1998, IIASA's Environmentally Compatible Energy Strategies (ECS) Program completed a five-year joint study with the World Energy Council (WEC) using the MESSAGE model (Nakićenović *et al.*, 1998). The study analysed six alternative global energy scenarios extending to 2100. The scenarios cover a wide range: from a strong expansion of coal production to strict limits, from a phase-out of nuclear energy to a substantial increase and from carbon emissions in 2100 that are only one-third of 1990 levels to increases by more than a factor of three.

Intergovernmental Panel on Climate Change (IPCC) Special Report on Emissions Scenarios (SRES)

A set of scenarios were developed for SRES which illustrate that similar future GHG emissions can result from very different socioeconomic developments, and that similar developments in driving forces can nonetheless result in widely different future emissions (Nakićenović and Swart, 2000). Thus the SRES reveals many continuing uncertainties that climate research and policy analysis must take into account. Results are presented principally in terms of projected emissions of CO_2 and other gases related to global warming. Six models from around the world were used to develop the new SRES scenarios, including the MESSAGE. Among other contributions, ECS provided nine of the 40 SRES scenarios.

IPCC Third Assessment Report (TAR)

The MESSAGE model was also among the nine international modelling frameworks used for the development of a number of GHG-mitigation scenarios for the IPCC's Third Assessment Report (Metz *et al.*, 2001). The mitigation scenarios are based upon socioeconomic and technological developments depicted by the SRES baseline scenarios and serve as the basis for the identification of robust GHG-mitigation strategies. The analysis with the MESSAGE model covers a range of stabilization levels for atmospheric CO_2 (450, 550, 650 and 750ppmv) and a range of different baselines including the A1, A2 and B2 families from SRES. Among other things, an assessment of the impact of alternative stabilization targets on technology diffusion, economic development and the corresponding energy and carbon prices was performed.

NOTES

1. A brief description of these applications is found in section 5 below.
2. Note that the energy statistics for the year 2000 were not yet available during the period when the scenarios were developed (1997–2000). Hence, in some cases, the reported figures for the year 2000 might deviate slightly from actual statistics.
3. This 'standard' refers to studies by IIASA-ECS over at least ten years.
4. In IPCC-SRES, 'ALM' is used to represent a ROW-equivalent region including all developing countries in Africa, Latin America and the Middle East.
5. It has become conventional to use the term 'resources' for depletable and 'sources' for renewable forms of primary energy (World Energy Conference, 1986).
6. MESSAGE also includes the possibility of defining demand elasticities directly by piecewise linear interpolation without requiring iterations with MACRO (see the description on page 182), but, for the formulation of the mitigation scenarios, the use of MESSAGE-MACRO was preferred.
7. The MPS file format is a standard format for the preparation of input data for solvers of Linear Programming models. The format definition is beyond the scope of this description.
8. Instead of using the decimal notation {0, 1, 2, . . ., 10, 11, . . .} for numbering of the time periods, a predefined sequence of (ASCII) characters {0, a, b, c, . . ., z, A, B, C, . . .} is used. The reason is that the identifier taken for the numbering in a variable name needs to be only one character long.
9. For a definition of all level identifiers, see pages 169–73.
10. We therefore sometimes speak of *technology v* if the meaning of this abbreviation appears clear.
11. Leakage during the transportation and associated emissions are dealt with separately via emissions coefficients.
12. Note, however, that in our scenarios no load regions are used, and hence this term is omitted.
13. Again, readers not interested in model features that were not used for the description of the scenarios in this book can safely skip this and the following subsection.
14. MESSAGE also has the capability to use more than one objective function at a time. See the description in the next subsection.
15. This is no more restrictive than a fixed ratio between the angular movements of the steering wheel of a car and the steered wheels. In both cases, the steering matters more than the mechanism that determines how the direction is changed.

REFERENCES

Grauer, M., A. Lewandowski and L. Schrattenholzer (1982), 'Use of the reference level approach for the generation of efficient energy supply strategies', WP-82-19, IIASA, Laxenburg, Austria.

Gritsevskyi, A. (1996), 'Scenario Generatory', Internal Report, International Institute for Applied Systems Analysis, IIASA, Laxenburg, Austria.

Manne, A.S. and R. Richels (2003), 'MERGE: A model for evaluating the regional and global effects of GHG reduction policies' (http://www.stanford.edu/group/MERGE/).

Messner, S. and L. Schrattenholzer (2000), 'MESSAGE-MACRO: Linking an energy supply model with a microeconomic module and solving it iteratively', *Energy*, **25**, 267–82.

Metz, B., O. Davidson, R. Swart and J. Pan (eds) (2001), *Climate Change 2001: Mitigation*, Contribution of Working Group III to the Third Assessment Report of the Intergovernmental Panel on Climate Change, Cambridge: Cambridge University Press.

Murtagh, B.A. and M.A. Saunders (1983), 'MINOS 5.0 USER'S Guide', Technical Report SOL 83-20, Systems Optimization Laboratory, Department of Operations Research, Stanford University.

Nakićenović, N., A. Grübler and A. McDonald (eds) (1998), *Global Energy Perspectives*, Cambridge: Cambridge University Press.

Nakićenović, N., J. Alcamo, G. Davis, B. de Vries, J. Fenhann, S. Gaffin, K. Gregory, A. Gruebler, T.Y. Jung, T. Kram, E.L. La Rovere, L. Michaelis, S. Mori, T. Morita, W. Pepper, H. Pitcher, L. Price, K. Riahi, R.A. Roehrl, H.-H. Rogner, A. Sankovski, M. Schlesinger, P. Shukla, S. Smith, R. Swart, S. van Rooijen, N. Victor and Z. Dadi (2000), *Special Report on Emissions Scenarios* (SRES), A Special Report of Working Group III of the Intergovernmental Panel on Climate Change, Cambridge: Cambridge University Press.

World Energy Conference (WEC) (1986), *Energy Terminology*, 2nd edn, Oxford: Pergamon Press.

Index